Free Radical Medicine and Biology

Edited by Kusal Das, Swastika Das, Mallanagouda Shivanagouda Biradar, Varaprasad Bobbarala and S. Subba Tata

Published in London, United Kingdom

IntechOpen

Supporting open minds since 2005

Free Radical Medicine and Biology
http://dx.doi.org/10.5772/intechopen.77829
Edited by Kusal Das, Swastika Das, Mallanagouda Shivanagouda Biradar, Varaprasad Bobbarala and S. Subba Tata

Part of IntechOpen Book Series: Physiology, Volume 10
Book Series Editor: Angel Catala

Contributors
Tae Gyu Choi, Sung Soo Kim, Yi Fang, Rongzhen Zhong, Kusal K. Das, Swastika Das, Shrilaxmi Bagali, Ishwar Bagoji, Mallanagouda Shivanagouda Biradar, Elisabetta Straface, Sandra Claro, Alice Ferreira, Maria Etsuko Myamoto Oshiro, Ziad Moussa, Zaher M. A. Judeh, Saleh A. Ahmed, Sayandeep K. Das, Aravind V. Patil, Rosa Vona, Donatella Pietraforte, Lucrezia Gambardella, Alessandra Marchesi, Isabella Tarissi de Jacobis, Alberto Villani, Domenico Del Principe

First published in London, United Kingdom, 2020 by IntechOpen
IntechOpen is the global imprint of INTECHOPEN LIMITED, registered in England and Wales, registration number: 11086078, 7th floor, 10 Lower Thames Street, London, EC3R 6AF, United Kingdom
Printed in Croatia

British Library Cataloguing-in-Publication Data
A catalogue record for this book is available from the British Library

Additional hard and PDF copies can be obtained from orders@intechopen.com

Free Radical Medicine and Biology
Edited by Kusal Das, Swastika Das, Mallanagouda Shivanagouda Biradar, Varaprasad Bobbarala and S. Subba Tata
p. cm.
Print ISBN 978-1-78985-143-4
Online ISBN 978-1-78985-144-1
eBook (PDF) ISBN 978-1-83962-595-4
ISSN 2631-8261

For EU product safety concerns:
IN TECH d.o.o., Prolaz Marije Krucifikse Kozulić 3, 51000 Rijeka, Croatia, info@intechopen.com or visit our website at intechopen.com.

IntechOpen Book Series
Physiology
Volume 10

Kusal K. Das is Professor of Physiology and director at the Laboratory of Vascular Physiology and Medicine, Shri B. M. Patil Medical College, BLDE (Deemed to be University), Vijayapur, Karnataka, India. Professor Das is also coordinator of environmental health under UNESCO Chair Life Sciences, Yerevan, Armenia. He was Visiting Professor of Medicine at the School of Medicine, University of Leeds (2014–2016). Prof. Das obtained a PhD and a master's degree in Physiology from the University of Calcutta. His research focuses on experimental hypoxia and cell signaling mechanisms during heavy metal-induced stress and alteration of vascular integrity in both experimental and human models. He invented a modified method of determination of serum vitamin E. He is a fellow in the Royal Society of Biology, UK. He has published more than 100 Scopus papers and authored five books and numerous book chapters.

Swastika N. Das is Professor of Chemistry at the V. P. Dr. P. G. Halakatti College of Engineering and Technology, BLDE (Deemed to be University), Vijayapura, Karnataka, India. She obtained an MSc, MPhil, and PhD in Chemistry from Sambalpur University, Odisha, India. Her areas of research interest are medicinal chemistry, chemical kinetics, and free radical chemistry. She is a member of the investigators who invented a new modified method of estimation of serum vitamin E. She has authored numerous publications including book chapters and is a mentor of doctoral curriculum at her university.

M. S. Biradar is Vice Chancellor and Professor of Medicine of BLDE (Deemed to be University), Vijayapura, Karnataka, India. He obtained his MD with a gold medal in General Medicine and has devoted himself to medical teaching, research, and administrations. He has also immensely contributed to medical research on vascular medicine, which is reflected by his numerous publications including books and book chapters. Professor Biradar was also Visiting Professor at Tulane University School of Medicine, New Orleans, USA.

Varaprasad Bobbarala has a doctorate from Andhra University with a specialization in Biochemistry, Medicinal Chemistry, and Microbiology. He is currently editor-in-chief, associate editor, editorial board member as well as reviewer of dozens of high-impact international periodicals. He has authored/co-authored research and review articles in numerous peer-reviewed national and international journals in various subjects related to biomedicine, pharmacy, and microbiology. Dr. Varaprasad previously served as the Chief Scientist of Research and Development at Krisani Innovations Pvt. Ltd., before his current role as the Chief Scientist and Director of Adhya Biosciences Pvt.

Ltd., India. He is currently working in the area of clinical diagnostics, antimicrobial resistance, drug discovery, production of commercially important chemicals by biotechnology routes, isolation of bio-active metabolites, and bio-efficacy studies.

Dr. S. Subba Tata has an MSc in Biotechnology), MSc in Botany, MTech in Biotechnology, and a PhD in Botany. He is an assistant professor in the Department of Botany, Andhra University, Visakhapatnam, India. He is qualified in CSIR-NET. APSET, ASRB-NET, and GATE competitive examinations in life sciences. He has authored two books and published twenty-two research papers in national and international journals. He participated and presented at thirty seminar/symposia organized in India.

Editors of Volume 10:
Kusal K.Das,PhD,FRSB
Laboratory of Vascular Physiology and Medicine
Shri B.M.Patil Medical College, Hospital and Research Centre
BLDE (Deemed to be University)
Vijayapur-586103, Karnataka, India

Swastika Das,M.Sc,M.Phil,PhD
BLDEA's V.P.Dr/P.G.Halakatti, College of Engineering & Technology, Vijayapur, Karnataka, India

Mallanagoud S.Biradar,MD
BLDE(Deemed to be University), Vijayapur-586103, Karnataka, India

Varaprasad Bobbarala PhD
Adhya Biosciences, India

S. Subba Tata
Andhra University, India

Book Series Editor: Angel Catala
National University of La Plata, Argentina

Scope of the Series

Modern physiology requires a comprehensive understanding of the integration of tissues and organs throughout the mammalian body, including the expression, structure, and function of molecular and cellular components. While a daunting task, learning is facilitated by our identification of common, effective signaling pathways employed by nature to sustain life. As a main example, the cellular interplay between intracellular Ca2 increases and changes in plasma membrane potential is integral to coordinating blood flow, governing the exocytosis of neurotransmitters and modulating genetic expression. Further, in this manner, understanding the systemic interplay between the cardiovascular and nervous systems has now become more important than ever as human populations age and mechanisms of cellular oxidative signaling are utilized for sustaining life. Altogether, physiological research enables our identification of clear and precise points of transition from health to development of multi-morbidity during the inevitable aging process (e.g., diabetes, hypertension, chronic kidney disease, heart failure, age-related macular degeneration; cancer). With consideration of all organ systems (e.g., brain, heart, lung, liver; gut, kidney, eye) and the interactions thereof, this Physiology Series will address aims of resolve (1) Aging physiology and progress of chronic diseases (2) Examination of key cellular pathways as they relate to calcium, oxidative stress, and electrical signaling & (3) how changes in plasma membrane produced by lipid peroxidation products affects aging physiology.

Contents

Preface

Free radicals are highly unstable and react quickly with the nearest stable molecules, "stealing" their electrons to gain stability and, in this way, initiating a chain reaction and propagation until finally attaching the vital molecules of cell and cell membrane thus disrupting their integrity. Free radical formation occurs continuously in the cells as a consequence of both enzymatic and non-enzymatic reactions. Enzymatic reactions, which serve as the source of free radicals, include those involved in the respiratory chain, phagocytosis, prostaglandin synthesis, and the cytochrome P-450 system.

Oxidative stress, arising as a result of an imbalance between free radical production and antioxidant defenses, is associated with damage to a wide range of molecular species including lipids, proteins, and nucleic acids. Short-term oxidative stress may occur in tissues injured by trauma, infection, heat injury, hypertoxia, toxins, and excessive exercise. These injured tissues produce increased radical-generating enzymes (e.g., xanthine oxidase, lipogenase, cyclooxygenase), activation of phagocytes, release of free iron, copper ions, or a disruption of the electron transport chains of oxidative phosphorylation, producing excess reactive oxygen species (ROS).

This book highlights various issues of free radical biology from the perspective of antioxidant defense mechanisms. It is dived into two sections: "Redox Biology and Medicine" and "Antioxidants." The first chapter in Section 1 is "Introductory Chapter: Free Radical Biology in Metal Toxicities—Role of Antioxidants." This chapter explains how oxygen metabolism induces overproduction of free radicals due to altered pathophysiology in the system, and how inadequate antioxidants in the body induces impaired body defense systems and inability to fight against pollutants, particularly heavy metals. The chapter concludes that mode of action of free radical-generated oxidative stress typically starts with the reaction of heavy metals with target molecules and ends with toxic manifestations. These processes of oxidative stress and oxidant and antioxidant imbalances play a key role in free radical biology. Chapter 2, "Physiological Functions of Mitochondrial Reactive Oxygen Species," discusses the mechanisms regulating mitochondrial physiology and homeostasis. Production of mitochondrial reactive species (mROS) may have significant potential for the development of novel therapeutic strategies for the treatment of a wide range of human pathologies. Chapter 3, "Role of Oxidative Stress in the Cardiovascular Complications of Kawasaki Disease" elaboartes on a complex framework of events contributing to the etiology of Kawasaki disease (KD). This chapter provides new lines of evidence supporting the hypothesis that systemic oxidative stress together with premature aging of red blood cells and platelets could play a critical role in the cardiovascular risk observed in patients with KD. Chapter 4, "Radiation-Generated ROS Induce Apoptosis via Mitochondrial" describes the role of mitochondria and characterizes the proteins involved in Ionizing Radiation(IR)-induced apoptosis. Further, the chapter concludes that IR triggers the activation of anti-apoptotic proteins and enhances the risk of a second type of cancer in patients undergoing radiotherapy. In addition to increasing the radioresistance of cells, anti-apoptotic proteins can also stimulate uncontrolled cell proliferation that culminates in mutagenesis.

Section 2, "Antioxidants," includes two chapters. Chapter 5, "Effects of Oxidative Stress on Spermatozoa and Male Infertility," examines the delicate balance between ROS necessary for physiological activity and antioxidants to protect from cellular oxidative injury, both of which are essential for fertility. The chapter authors suggest that determination of the appropriate antioxidant compounds as well as their dosages, whether used in clinical practice or cryopreservation, need to be explored.

Finally, Chapter 6, "Nonenzymatic Exogenous and Endogenous Antioxidants" explains the chemical structure and mechanism of action of the most important non-enzymatic small exogenous and endogenous organic molecules that act as antioxidants. The chapter authors further conclude that the regulation of gene expression by employing oxidants and antioxidants represents a novel approach with promising therapeutic implications.

This book enriches the understanding of oxidative stress biology with reference to some specific antioxidant defense mechanisms. It is designed for students, researchers, and general readers with an interest in the subject matter.

We acknowledge the support of Dr. Shrilaxmi Bagali, Dr. Sayandeep K. Das, Dr. Aravind V. Patil, and Dr. Ishwar B. Bagoji for their help during the publication of this book.

Kusal K. Das, PhD, FRSB
Professor of Physiology,
Laboratory of Vascular Physiology and Medicine,
Shri B.M. Patil Medical College, Hospital and Research Centre,
BLDE (Deemed to be University),
Vijayapura, Karnataka, India

Swastika Das, M.Sc, M.Phil, PhD
Professor of Chemistry,
BLDEA's V.P.Dr/P.G.Halakatti,
College of Engineering and Technology,
Vijayapura, Karnataka, India

Mallanagoud S. Biradar, MD
Professor of Medicine,
BLDE (Deemed to be University),
Vijayapura, Karnataka, India

Dr. Varaprasad Bobbarala, PhD
Chief Scientist and Director,
Adhya Biosciences Pvt. Ltd.,
India

Dr. S. Subba Tata
Assistant Professor,
Department of Botany,
Andhra University,
India

Section 1

Redox Biology and Medicine

Chapter 1

Introductory Chapter: Free Radical Biology in Metal Toxicities—Role of Antioxidants

Swastika Das, Shrilaxmi Bagali, Sayandeep K. Das, Aravind V. Patil, Ishwar B. Bagoji, Kusal K. Das and Mallanagouda S. Biradar

1. Introduction

Industrial revolution imparts a high level of metal contamination in this world [1–3]. Although technology advances to control pollution, it fails to check environmental heavy metal pollution. As metals used not always leads to biodegradation, hence, these heavy metals circulate in human life chains and accumulate in living organisms [4]. Some metals like zinc and iron are physiologically essential, but they may also alter the function of organisms when the exposure dose exceeds a critical threshold, which is species specific, and can depend on multiple factors including age, sex, and functional norms of physiological system [5]. Rodents are as sensitive and near similar to human monitoring system against metal toxicity studies [6].

In the last few decades there has been a tremendous interest on oxygen-free radicals, more generally known as "reactive oxygen species," (ROS) and of "reactive nitrogen species" (RNS) in experimental and clinical studies on various environmental pollutants [6]. ROS and RNS are found to be generated during irradiation or metal catalyzed reaction or its presence in atmosphere as pollutants or simply generates due to inflammation in the physiological system. Some examples of free radicals are hydrogen peroxide, singlet oxygen, hypochlorous acid, superoxide radical, hydroxyl radical, and nitric oxide [7].

It has generated interest that oxygen metabolism induces over production of free radicals due to altered pathophysiology in the system. Another hypothesis is that inadequate antioxidants in the body also impairs body defense system and ability to fight against pollutants especially heavy metal pollutants. Antioxidants can prevent cellular damage by interacting with free radicals and terminating chain reaction process [8, 9].

2. Metal toxicities

Metals play an important role by conjugating at the active sites of enzymes and participate directly in catalysis, stabilize macromolecular structures of proteins and nucleic acids, thereby affecting structural and functional integration.

The molecular mechanisms including enzymatic functions of metal induced bio-toxicities have been established. Possible recognition of essential biological roles of metals, of course, in no way obviates the primary objective of ecological and toxicological investigation, i.e., to eliminate the hazards created by metals. In this regard, Bertrand's early enunciation of the necessity to consider physiological and toxicological effects of metals as a biological continuum is important [10]. Bertrand further emphasized that, metals induce a double humped, biphasic dose response curve, which allows a gross division into two general regions.

i. Potentially, every element has a biological function which can be assessed properly only against a background of deficiency state.

ii. Potentially, every element is toxic when presented to an organism in high enough concentration.

The toxicity of a metal or its compounds in a biological system is influenced by a number of factors like:

i. The intrinsic toxicity of metal

ii. The dose of metal

iii. The combining capacity of metal

iv. The capacity of biological system to absorb and transport the metal to the target organ most susceptible to the metal intoxication

v. The capacity of the metal to transform to a less toxic or a more toxic form at the target organ or during transfer

vi. The ability of the metal to bind to essential macromolecules

vii. The homeostatic mechanism of the organism to either excrete or sequester the metal

Excess doses of some non-toxic metal compounds may interfere with normal cellular or physiological process by non-specific activity such as changing the osmotic pressure and pH or physically changing the microenvironment of the GI tract. The defensive homeostatic mechanism of cells and tissues combat metal intoxication either by sequestering the metal in a harmless way or by enhanced excretion of the toxic metal [11]. Cellular injury from toxic metals may occur by a number of diverse molecular mechanisms and at many levels of biological organization within a given target organ or cell population. It has become increasingly evident that the toxic potential of metals such as nickel, cadmium, mercury, lead is highly dependent on their intracellular bioavailability. Apart from these, there are several interactive factors that are capable of influencing the toxic effects of these heavy metals. Nutritional status, the presence or absence of other essential metals, contribution of proper antioxidants may greatly alter the distribution of metals within intracellular compartments. Normally it has been found that occupational metal exposed toxicities in developed countries are in decreasing trend but in third world countries the toxicities from metal exposure is still serious issues which make an impact on the health of the occupationally exposed people with impairments of neurological, reproductive, immunological and cardiovascular functions [12]. The European Commission initiated an action program concerning the environmental protection from heavy metals exposure [13].

3. Free radicals biology and metals

Oxidation occurs when free radicals attack biological molecules, removing an electron. Under certain conditions, unsaturated fatty acids can undergo oxidation, known as lipid peroxidation which sets off a chain reaction that generates large number of free radicals, which are both cytotoxic and genotoxic capable of altering DNA functions. The mechanism of heavy metal toxicity through electron transfer most often involves the cross linking of the sulfhydryl groups of proteins. Free radicals can also be generated directly from molecular oxygen in a two-step process to produce superoxide anion. In the continued presence of heavy metals, the superoxide anions formed can then combine with protons in the dismutation reaction, generating hydrogen peroxide (H_2O_2) in the process. Superoxide anions can also produce highly toxic hydroxyl radicals. Suitable mechanisms have evolved so that the steady-state concentrations of potentially toxic oxygen-derived free radicals are kept in check under normal physiological conditions by the body's intrinsic antioxidant defense system. Nevertheless, the enhanced generation of these ROS can overwhelm the intrinsic defenses of the cell, resulting in a condition known as oxidative stress [14]. Heavy metal can produce oxidative stress; therefore, it was conceivable that reactive oxygen species (ROS) may trigger signaling pathways resulting in the activation of the hypoxia-inducible factor (HIF)-1 transcription factor and up-regulation of hypoxia-related genes. The activity of the HIF-1 transcription factor as assessed in transient transfection assays was stimulated by heavy metals but this activation was not diminished when oxidative stress was attenuated nor was HIF-dependent transcription enhanced by hydrogen peroxide. It was reported that ROS are produced during the exposure of cells to metals that mimic hypoxia, but the formation of ROS was not involved in the activation of HIF-1-dependent genes [15]. One explanation of the heavy metal-induced activation of the HIF-1 transcription factor is based on the assumption that it replaces iron in the oxygen carrier, Fe(II)-hybrid hemoglobin. Substitution of iron by other heavy metals switch signal to permanent hypoxia, which in turn activates the HIF-1 factor [16]. The pretreatment of human blood lymphocytes with either CAT (a H_2O_2 scavenger), or SOD (a scavenger of O_2^- radical) significantly reduced markers of heavy metal- induced genetic and cellular damage. Glutathione depletion, a marker of oxidative stress, was found in human alveolar epithelial Type II-like cell line after treatment with heavy metal containing ultra-fine metal dust [17]. After controlling for confounders, plasma lipid peroxidation levels were significantly increased and erythrocyte antioxidants were significantly decreased in metal exposed experimental animals as compared with controls [18].

4. Role of antioxidant on metal toxicity

Free radicals are reactive chemical species that contain one or more unpaired electrons e.g. hydrogen peroxide, singlet oxygen, hypochlorous acid, superoxide radical, hydroxyl radical, and nitric oxide. As heavy metals generate free radicals hence it must be quenched by an antioxidant otherwise these free radicals will react with membrane lipid, protein, carbohydrate and nucleic acid molecules and change their functional moiety in cellular system. It has been found that human nuclear DNA receives approximately 10,000 oxidative 'hits' every day. It clearly reflects that each of the cells is under firing from ROS and the situation becomes worst if the cell is targeted by heavy metals. The antioxidant defense system against metal toxicities are fundamentally superoxide dismutase, glutathione, peroxidases, and catalase besides Fe and Cu binding proteins like albumin, transferrin, lactoferrin, haptoglobin, uric acid, bilirubin and carotenoids. Further antioxidant vitamins like vitamin C, vitamin E, Vitamin B12 are also considered as protective agents against metal toxicities. Some endogenous and exogenous polyphenolic compounds like flavonoids and ligands are also found to be protective against metal toxicities as antioxidants. Finally, there are specific nuclear repair enzymes, proteases, and other enzymes that constantly target oxidized molecules for catabolism [19]. Antioxidants are intimately capable of protecting cellular damage by interfering with ROS and stop the free radical due to metal induced chain reactions.

The success of an antioxidant against metal induced oxidative stress depends on its capability to (i) quenching free radicals (ii) chelating redox metals (iii) regenerate some more antioxidants within "antioxidant network", (iv) successfully induce cell signaling to express adaptive genes, (v) readily absorption capability, (vi) must have adequate concentration in tissue and biofluid and (vii) capability to act on both membrane and aqueous areas.

Regarding antioxidant supplementation against metal induced oxidative stress one must remember that higher doses of supplementary antioxidants do not always offer protection against free radicals. It is widely accepted that in a healthy organism there exists a balance between oxidants and various antioxidants. High levels of antioxidants may also disturb oxidant and antioxidant balance with unpredictable and unexpected consequences.

5. Conclusion

The steps of metal toxicity are as following: liberation of toxic metal > reaction with target molecules > cellular dysfunction > respond to reaction (repair) or (disrepair) > developmental toxicity. Mode of action typically starts with the reaction of metals with target molecules and ends with toxic manifestations and entire these process oxidative stress and oxidant and antioxidant imbalances play a key role.

Acknowledgements

The corresponding author greatly acknowledges Vision Group of Science and Technology, Government of Karnataka (VGSTKFIST/ 1230/2015-2016 Dated 22/6/2016) for financial assistance under K-FIST, Level.

Author details

Swastika Das[1], Shrilaxmi Bagali[2], Sayandeep K. Das[3], Aravind V. Patil[4], Ishwar B. Bagoji[5], Kusal K. Das[2*] and Mallanagouda S. Biradar[6]

1 Department of Chemistry, BLDEA's V.P. Dr. P.G. Halakatti College of Engineering and Technology, Vijayapura, Karnataka, India

2 Laboratory of Vascular Physiology and Medicine, Department of Physiology, Shri B.M. Patil Medical College, Hospital and Research Centre, BLDE (Deemed to be University), Vijayapura, Karnataka, India

3 Intern, Shri B.M. Patil Medical College, Hospital and Research Centre, BLDE (Deemed to be University), Vijayapura, Karnataka, India

4 Department of Surgery, Shri B.M. Patil Medical College, Hospital and Research Centre, BLDE (Deemed to be University), Vijayapura, Karnataka, India

5 Department of Anatomy, Shri B.M. Patil Medical College, Hospital and Research Centre, BLDE (Deemed to be University), Vijayapura, Karnataka, India

6 Department of Medicine, Shri B.M. Patil Medical College, Hospital and Research Centre, BLDE (Deemed to be University), Vijayapura, Karnataka, India

*Address all correspondence to: kusaldas@yahoo.com; kusaldas@bldedu.ac.in

References

[1] Goyer R. Toxic effects of metals. In: Amdur MO, Doull JD, Klaassen CD, editors. Casarett and Doull's Toxicology. 4th ed. New York: Pergamon Press; 1991. pp. 623-680

[2] Sunderman FW Jr, Oskarsson A. Nickel. In: Merian E, editor. Metals and their Compounds in the Environment. New York: VCH Verlagsgesellschaft; 1991. pp. 1101-1126

[3] Ercal N, Gurer-Orhan H, Aykin-Burns N. Toxic metals and oxidative stress part I: Mechanisms involved in metal induced oxidative damage. Current Topics in Medicinal Chemistry. 2001;**1**:529-539

[4] Merian E. Metals and Their Compounds in the Environment: Occurrence, Analysis and Biological Relevance. Weinham: VCH Verlagsgesellschaft GmbH; 1991. pp. 134-187

[5] Shore RF, Rattner BA, editors. Ecotoxicology of Wild Animals. Chichester, United Kingdom: John Wiley & Sons, Ltd.; 2001. p. 730

[6] Salnikow K, Donald SP, Bruick RK, Zhitkovich A, Phang JM, Kasprzak KS. Depletion of intracellular ascorbate by the carcinogenic metal nickel and cobalt results in the induction of hypoxic stress. The Journal of Biological Chemistry. 2004;**279**:40337-40344

[7] Halliwell B, Whiteman M. Measuring reactive species and oxidative damage in vivo and in cell culture: How should you do it and what do the results mean? British Journal of Pharmacology. 2004;**142**(2):231-255

[8] Cadenas E. Biochemistry of oxygen toxicity. Annual Review of Biochemistry. 1989;**58**:79-110

[9] Das KK, Das S, Ambekar JG. Chapter 11: Hypoxia and oxidative stress: Cell signaling mechanisms and protective role of vitamin C and cilnidipine. In: Catala A, editor. Lipid Peroxidation: Inhibition, Effects and Mechanisms. NY: Nova Science Publishers; 2017

[10] Bertrand G. On the role of trace substances in agriculture. Eighth International Congress of Applied Chemistry. 1912;**28**:30-40

[11] Das KK, Reddy RC, Bagoji IB, Das S, Bagali S, Mullur L, et al. Primary concepts of nickel toxicity; an overview. Journal of Basic and Clinical Physiology and Pharmacology. 2018. DOI: 10.1515/jbcpp-2017-0171

[12] Alessio E, Balducci G, Lutman A, Mestroni G, Calligaris M, Atti WM, et al. Synthesis and characterization of two new classes of ruthenium(III)-sulfoxide complexes with nitrogen donor ligands (L): Na[trans-RuCl4(R2SO)(L)] and mer, cis-RuCl3(R2SO)(R2SO)(L). Inorganica Chimica Acta. 1993;**203**(2):205-217

[13] Das KK, Das SN, Dhundasi SA. Nickel, its adverse health effects & oxidative stress. The Indian Journal of Medical Research. 2008;**128**(4):412-425

[14] Das KK, Buchner V. Effect of nickel exposure on peripheral tissues: Role of oxidative stress in toxicity and possible protection by ascorbic acid. Reviews on Environmental Health. 2007;**22**(2):157-173

[15] Salnikow K, Costa M, Figg WD, Blagosklonny MV. Hyperinducibility of hypoxia-responsive genes without p53/p21-dependent checkpoint in aggressive prostate cancer. Cancer Research. 2000;**60**:5630-5634

[16] Lynn R, Talbot JA, Morgan DL. Differences in rat skeletal muscles after incline and decline running. Journal of Applied Physiology. 1998;**85**(1):98-104

[17] Chakrabarti SK, Bai C, Subramanian KS. DNA-protein crosslinks induced by nickel compounds in isolated rat renal cortical cells and its antagonism by specific amino acids and magnesium ion. Toxicology and Applied Pharmacology. 1999;**154**(3):245-255

[18] Das KK, Gupta AD, Dhundasi SA, Patil AM, Das SN, Ambekar JG. Protective role of L-ascorbic acid on antioxidant defense system in erythrocytes of albino rats exposed to nickel sulfate. Biometals. 2007;**20**(2):177-184

[19] Das KK, Jargar JG, Hattiwale SH, Yendigeri SM, Das S, Dhundasi SA. Serum vitamin E (alpha-Tocopherol) estimation. A potential biomarker of antioxidant status evaluation on heavy metal toxicities. Recent Patents on Biomarkers. 2013;**3**:36-43

Chapter 2

Physiological Functions of Mitochondrial Reactive Oxygen Species

Tae Gyu Choi and Sung Soo Kim

Abstract

Mitochondria are the major energy producers within a cell in the form of adenosine triphosphate by oxidative phosphorylation. Normal mitochondrial metabolism inevitably generates reactive oxygen species (ROS), which have been considered to solely cause cellular damage. Increase of oxidative stress has been linked to various pathologies. Thus, mitochondrial ROS (mROS) were basically proposed as byproducts of oxidative metabolism, which undergo normalized by antioxidant enzymes. However, the mROS have extensively been esteemed to function as signalling molecules to regulate a wide variety of physiology. These phenomena are indeed dependent on mitochondrial redox status, which is dynamically altered under different physiological and pathological conditions. The oxidative stress is incurred by which the redox status is inclined to exceeded oxidation or reduction. Here, we attempt to integrate the recent advances in our understanding of the physiological functions of mROS.

Keywords: mitochondrial ROS, oxidative stress, oxidative metabolism, redox signaling, mitochondrial physiology

1. Introduction

Mitochondria are double-membrane-bound cellular organelles found in most eukaryotic organisms. The number of mitochondria in cell differs widely according to organisms, tissues and cell types, which is determined by the energy demand. Mitochondria occupy around 40% of the cytoplasm in heart muscle cells and 20–25% with ~2000 per cell in liver cells. Mitochondria, as the power plants of the cell, mainly generate energy in forms of adenosine triphosphates (ATPs) by oxidative phosphorylation (OXPHOS) during glucose metabolism [1, 2]. The OXPHOS is coupled with mitochondrial respiration in which mitochondrial transmembrane potential (MMP, $\Delta\Psi m$) is generated by pumping the protons via mitochondrial complexes I, III and IV of the electron transport chain (ETC) [3].

Molecular oxygen (O_2) is essential for the mitochondrial bioenergetic metabolism, which functions as the final electron acceptor for cytochrome *c* oxidase (complex IV) in the respiratory ETC that catalyses the four-electron reduction of O_2 to H_2O. Mitochondria are an important source of reactive oxygen species (ROS) within most mammalian cells [4, 5]; mitochondrial ROS (mROS) are basically produced as byproducts of this bioenergetic metabolism during the OXPHOS [6]. Electron leaks at complex I and III from ETC lead to forming partially reduced and highly reactive metabolites of O_2, including superoxide anion ($O_2^{\cdot-}$) and hydrogen peroxide (H_2O_2), formed by one- and two-electron reductions of O_2, respectively [7]. In the presence of transition metal ions, the more reactive hydroxyl radical ($OH^{\cdot}$) is formed. The $O_2^{\cdot-}$ is rapidly dismutated to H_2O_2 by two dismutases including Cu/Zn-superoxide dismutase (Cu/ZnSOD) in mitochondrial intermembrane space and manganese-dependent superoxide dismutase (MnSOD) in mitochondrial matrix. Unless the dismutation of $O_2^{\cdot-}$ is catalyzed into H_2O_2, the radical oxidant promotes DNA damage, protein oxidation and lipid peroxidation in many types of cells. H_2O_2 is also cell damaging molecule to be degraded to water by catalase [8]. Although the $O_2^{\cdot-}$ generation by respiratory complexes is a well-established phenomenon, it is still poorly understood in mechanism [9].

Mitochondria have been implicated in the regulation of a number of physiological and pathological processes, including proliferation, differentiation, programmed cell death, innate immunity, autophagy, redox signalling, calcium homeostasis, hypoxic stress responses and stem cell reprogramming [10–16]. The mROS production contributes to mitochondrial damage in a range of pathologies, which is also is closely related to redox signalling in the cell [4, 17]. However, accumulating evidences show that mROS are not only deleterious molecules derived from the cellular metabolism but also indispensable participants in diverse cellular signalling and regulations [18–20].

In this chapter, we briefly summarize recent developments in our understanding of the involvement of mROS as signalling mediators in redox biology, rather than pathological stress, underlying physiological conditions.

2. Mitochondrial physiology and ROS production

Mitochondria, cellular organelles in cells of eukaryotic organisms, have a primarily role in the process of pyruvate breakdown and ATP synthesis, generating water and carbon dioxide (CO_2) as the end products via aerobic respiration [21]. Mitochondria turned into driving forces in biological evolution after the symbiotic engulfment of aerobic α-proteobacteria by a precursor of the eukaryotic cells around 2 billion years ago [22, 23]. Although mitochondria have preserved the double-membrane shape and ATP production characters of their ancestors, they have attained numerous additional functions in the cell, dramatically altering their structure and composition [24]. Most part of the bacterial genome was rapidly lost or transferred to the nuclear DNA [25]. Mammalian mitochondrial genome is transmitted solely through the female germ line [26]. Human mitochondrial DNA (mtDNA) is a double-stranded, circular molecule of 16,569 bp and contains 37 genes coding for two ribosomal RNAs (rRNAs), 22 transfer RNAs (tRNAs) and 13 proteins [22]. As the major power plants, mitochondria constantly produce reactive radical oxidants as byproducts during OXPHOS. Thus, in response to the metabolic or environmental stresses, mitochondria have accomplished antioxidant defence system [27]. Mitochondria are also highly dynamic to maintain the functions, which form a tubular network that continually changes by fission and fusion [28]. In this section, we concisely discuss overall mitochondrial biology and the ROS generation.

2.1 Mitochondrial structure and genome

A mitochondrion comprises four subcompartments, the outer mitochondrial membrane (OMM) and inner mitochondrial membrane (IMM), and the two soluble compartments intermembrane space (IMS) and matrix, which are architecturally and functionally distinct. The OMM encloses the organelle, which has a protein-to-phospholipid ratio similar to that of the plasma membrane [29]. It contains large numbers of integral membrane protein, porin [30]. Voltage-dependent anion channel (VDAC) is a major trafficking protein that forms a beta barrel spanning the outer membrane, which transports nucleotides, ions and metabolites between cytosol and intermembrane space [31, 32]. The IMM is found inside of the OMM, which encloses mitochondrial matrix, extensively folded and compartmentalized [33]. The IMM is non-permeable to nucleotides, sugars and small ions; thus specific carrier proteins enable the molecules to transport across the membrane [34].

The mitochondrial respiratory complexes I–IV, in which electrochemical gradient is generated for OXPHOS to occur for ATP synthesis, are embedded in the IMM [22]. Mitochondria contain two aqueous compartments: the IMS and matrix. The IMS, existing between the OMM and IMM, relays molecular transport from cytosol to mitochondrial matrix or reversely [35]. The compositions of small molecules such as ions and sugars in IMS are chemically similar to those in cytosol, as OMM is selectively permeable to those molecules [36]. However, in case of large proteins, the specific signalling peptides are required to be transported across the OMM. Thus, the protein composition of the IMS is different to the protein composition of the cytosol (e.g. cytochrome c) [37]. The mitochondrial matrix, enclosed by IMM, contains mitochondrial DNA (mtDNA), RNA and proteins. Especially, a number of proteins in the matrix are involved in diverse biochemical processes such as tricarboxylic acid (TCA) cycle, fatty acid oxidation, amino acid degradation and mitochondrial dynamics (fission and fusion) [27, 38] (**Figure 1**).

2.2 Mitochondrial genome

Mitochondria contain their own genetic material (mtDNA), which is maternally inherited without DNA recombination and encodes 37 genes that participate in mitochondrial ATP synthesis. Thirteen genes of them are involved in OXPHOS, and the rest two rRNAs and 22 tRNAs. One human cell has hundreds to thousands of mtDNA copies [39, 40]. mtDNA has high rates of mutation and sequence evolution, and mutant and wild-type mtDNA are present in the cell at different proportions [41, 42]. The mtDNA mutations lead to abnormality in OXPHOS activity and ATP synthesis [43]. The mtDNA is exposed to OXPHOS-derived ROS without conventional histone proteins. Moreover, in the lacking repair mechanisms, mtDNA damage further amplifies during DNA replication [44]. Therefore, the mtDNA is susceptible to mutation

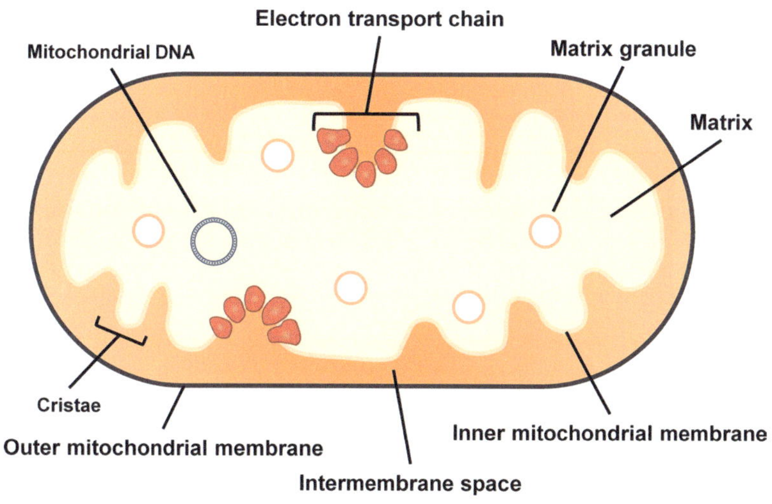

Figure 1.
Mitochondrial structure.

and damage. Therefore, mtDNA mutations and damage cause mitochondrial dysfunction, including ATP synthesis impediment, intracellular calcium level elevation, phospholipase activation and membrane phospholipids decomposition [45, 46].

2.3 Mitochondrial bioenergetics and dynamics

The most prominent roles of mitochondria are to produce the cellular energy in forms of ATPs via aerobic respiration and to regulate the cellular metabolism. Nutrients such as sugars (mostly glucose), lipids and amino acids are oxidized to primarily produce the energy [47]. Approximately 90% of cellular energy requirements are generated in mitochondria [48]. Glucoses and lipids are glycolysed into pyruvic and fatty acids, respectively, in the cytoplasm. Subsequently, these acids form acetyl coenzyme A (Acetyl CoA) via a series of catabolic reactions and then enter the TCA cycle in the mitochondrial matrix [49]. In the reaction of the TCA cycle, convert three equivalents of nicotinamide adenine dinucleotide (NAD^+) into three equivalents of reduced NAD^+ (NADH), one equivalent of flavin adenine dinucleotide (FAD) into one equivalent of $FADH_2$ and one equivalent each of guanosine diphosphate (GDP) and inorganic phosphate (P_i) into one equivalent of guanosine triphosphate (GTP). The NADH and $FADH_2$ are, in turn, used by the OXPHOS to generate ATPs. Thus, oxidation of nutrients provides electrons to the mitochondrial ETC in the form of NADH and $FADH_2$. The sequential transport of electrons from complex I or II to III and IV extrudes protons from the matrix to the IMS, generating an electrochemical gradient. In this process, the ETC requires two electron carriers: coenzyme Q_{10} (CoQ_{10}, also known as ubiquinone) and cytochrome c (Cytc) [50]. Along this electrochemical gradient, the protons flow through complex V (ATP synthase) on the IMM to return to the mitochondrial matrix. This reflux alters the conformation of complex V and drives the synthesis of ATP from ADP and P_i [47].

2.4 Mitochondrial dynamics (fission and fusion)

Mitochondria are highly dynamic and interacting organelles. Mitochondria are able to autonomously integrate (fusion) and divide (fission) by remodelling their morphology and moving along cytoskeletal tracks, in response to their metabolic or pathogenic conditions and cellular environment [29]. The mitochondrial lengths and networks are determined by the balance between fission and fusion rates [51]. Mitochondrial fission and fusion processes are mainly mediated by large guanosine triphosphatases (GTPases) in the dynamin family [51].

Mitochondrial fission requires the dividing of mitochondrial proteins and mtDNA thus that each daughter organelle normally functions without significant loss of soluble proteins from the mitochondrial matrix or intermembrane space [52]. Fission is required for the cellular distribution of mitochondria during cell division and embryonic growth [53]. Exceeded mitochondrial fission, not mutually balanced with fusion, leads to glucose oxidation, MMP decrease and hence the downregulation of ATP production [54]. The fission process is coordinated by a set of components in the cytosol, cytoskeleton, as well as mitochondria. Fission is mediated by a cytosolic dynamin family member, dynamin-related protein 1 (Drp1). Drp1 is recruited from the cytosol to form spirals around mitochondria and, subsequently, constricts the membranes at the fission site to split the mitochondrial cluster [29].

Mitochondrial fusion is mediated by a different set of dynamin-related GTPases. Mitochondrial outer membrane fusion is coordinated with inner membrane fusion. Three large GTPases are essential for mitochondrial fusion [55]. The mitofusins (Mfn1 and Mfn2) are transmembrane GTPases embedded in the OMM [56]. OPA1 is a dynamin-related GTPase associated with the IMM or IMS. Mitofusins and OPA1 physically interact to mechanistically mediate OMM and IMM fusion, respectively [29, 57, 58]. Mitochondrial fusion may increase to maximize the fidelity for OXPHOS in cellular energy demands [27].

2.5 Mitochondrial ROS production and antioxidant enzymes

Mitochondria are the major source of ROS generation [9]. In an organism, mitochondria utilize approximately 98% of the total amount of inhaled O_2, including 1–2% for ROS generation [59, 60]. Mitochondria actually produce ROS in a number of enzymatic reactions; the vast majority of the free radicals from the mitochondria are formed in the ETC during OXPHOS [61]. In the process of OXPHOS, electron leaks from the ETC combine with O_2 molecules to form ($O_2^{\cdot-}$). Mitochondrial $O_2^{\cdot-}$, primarily generated in complexes I and III, is catalysed by Cu/ZnSOD or Mn SOD to disproportionate into H_2O_2. Subsequently, H_2O_2 can be converted to $OH^{\cdot}$ by Fenton reaction. Mitochondrial $O_2^{\cdot-}$ can also bind with protons to form uncharged $HOO^{\cdot}$ radicals and subsequently react with unsaturated fatty acid of mitochondrial membrane lipids to produce lipid radicals. Mitochondrial nitric oxide (NO) interacts with $O_2^{\cdot-}$ to form reactive nitrogen oxide species (RNS) such as peroxynitrite ($ONOO^-$), which produce cellular dysfunction by S-nitrosylating proteins [62]. Mammalian cells have multiple enzymes to degrade H_2O_2, including peroxiredoxins (Prxs), glutathione peroxidases (Gpxs), thioredoxins (Trxs) and catalase. Mitochondrial H_2O_2 is primarily eliminated by the action of Gpx1, Gpx2 and Gpx4, Prx3 and Prx5 and Trx2 systems, which requires glutathione (GSH) [63–65]. Oxidized GSH (GSSG) is reduced to GSH by glutathione reductase (GR) activity [66]. Similarly, oxidized Trx2 is recycled by Trx reductase (TrxR). These H_2O_2 scavenging system ultimately depends on reduced nicotinamide adenine dinucleotide phosphate (NADPH) which is regenerated by three mitochondrial matrix-located enzymes: $NADP^+$-linked isocitrate dehydrogenase (IDH), malate dehydrogenase (MDH)

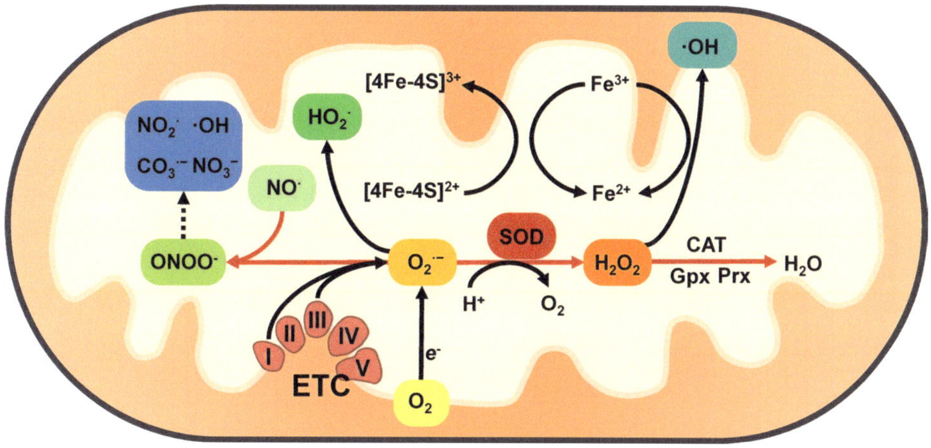

Figure 2.
Reactions and transformations of mitochondrial ROS. SOD enzymes catalyse the dismutation of superoxide ($O_2^{\cdot-}$), generating hydrogen peroxide (H_2O_2). The catalase (CAT), glutathione peroxidases (Gpxs) and peroxiredoxins (Prxs) convert H_2O_2 into water. H_2O_2 reacts with redox-active iron to generate the hydroxyl radical ($OH^{\cdot}$) through the Fenton reaction. The reaction between $O_2^{\cdot-}$ and nitric oxide ($NO^{\cdot}$) produces peroxynitrite ($ONOO^-$), whose decomposition in turn gives rise to some highly oxidizing intermediates including $NO_2^{\cdot}$, $OH^{\cdot}$, $CO_3^{\cdot-}$ as well as, finally, stable NO_3^-. Thus, increased $O_2^{\cdot-}$ levels can also reduce $NO^{\cdot}$ and generate $ONOO^-$ toxicity. $O_2^{\cdot-}$ by itself can reduce ferric iron (Fe^{3+}) to ferrous iron (Fe^{2+}) in iron-sulphur centres of proteins, leading to enzyme inactivation and concomitant loss of Fe^{2+} from the enzymes. The protonation of $O_2^{\cdot-}$ can form the more reactive hydroperoxyl radical ($HO_2^{\cdot}$).

and nicotinamide nucleotide transhydrogenase (NNT) [61]. Catalase catalyses the decomposition of hydrogen peroxide to water and oxygen, existing as a tetramer composed of four identical monomers, each of which contains a heme group at the active site. Catalase also requires NADPH as a reducing equivalent to prevent oxidative inactivation of the enzyme [67] (**Figure 2**).

3. Physiological functions of mitochondrial ROS in diverse cellular processes

mROS generation is a ubiquitous phenomenon during life of eukaryotic cells [68]. mROS-induced oxidative stress is considered a main contributor to the aetiology of both normal senescence and severe pathologies. Under normal physiological conditions, mROS emission is accounted for ~2% of the total O_2 consumption, of which the decomposition is well-controlled [2]. Accumulation of mROS, which is an imbalance of neutralization, induces deleterious consequences such as neurodegenerative disease [69], cardiovascular disease [70] and cancers [71]. However, depending on the cellular environment, antioxidant machinery-regulated oxidative stress could initiate diverse cellular responses, involved in cell protection, initiating coordinated activation of mitochondrial fission and autophagy to carry out clearance of abnormal mitochondria and cells, which are to protect spreading the damage to the adjacent cells [72, 73]. H_2O_2 is the primary molecule of mROS utilized for intracellular signalling, which selectively reacts with cysteine residues in redox-sensitive proteins, altering activities or conformations of the proteins to regulate signal transduction [74–76]. Mechanistically, H_2O_2 oxidizes thiol groups (SH) on cysteine residues to form sulphenic acid (SOH), which react with GSH to become glutathionylated (GSSG), with neighbouring thiols to form a disulphide bond (S-S) or with amides to form a sulphenyl amide (S-N) [77, 78]. In this section, we introduce the physiological roles and regulations of mROS in diverse cellular processes such as proliferation, differentiation, autophagy, immunity and aging (**Figure 3**).

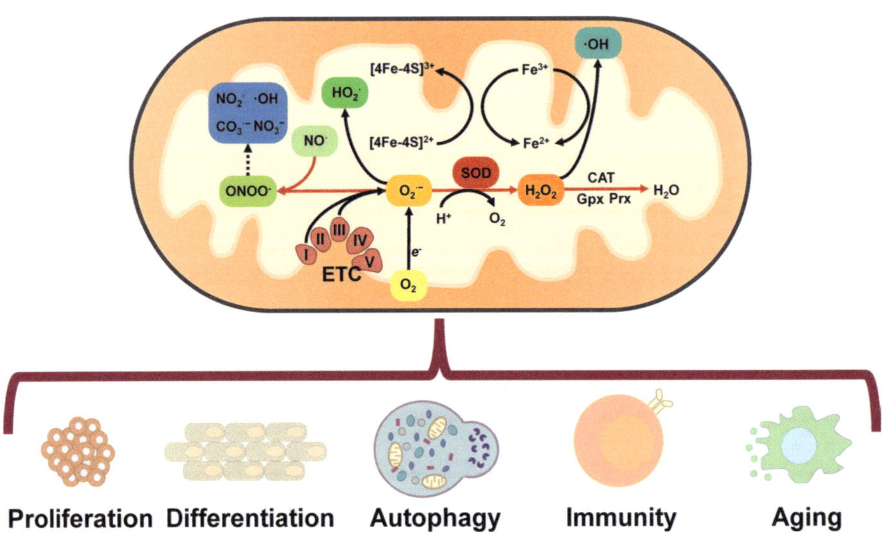

Figure 3.
Physiological regulation by mitochondrial ROS. mROS contribute to the various physiological cellular processes, including proliferation, differentiation autophagy, immunity and aging.

3.1 Proliferation

Accumulation of mitochondria-derived ROS enables to prompt cell proliferation inhibition and cellular senescence [79, 80]. However, the cells essentially utilize mROS for survival and growth via multiple mechanisms in diverse circumstances.

mROS regulate cell proliferation during hypoxia. Under the hypoxic condition (a low O_2 environment, generally 0.3–3% of O_2), the cells raise transcriptional and non-transcriptional responses to increase O_2 supply, simultaneously reducing O_2 consumption. These adaptations to hypoxia are enhanced by mROS. The hypoxia-inducible factors (HIFs) such as HIF1, HIF2 and HIF3 orchestrate the transcriptional response to the hypoxia, promoting erythropoietin (EPO) expression to increase erythropoiesis, vascular endothelial growth factor (VEGF) to promote blood vessel formation and glycolysis enzymes to retain ATP levels [81, 82]. HIFs are heterodimers consisting of two basic helix-loop-helix/PAS proteins: a stable β-subunit and one of three unstable labile α-subunits (HIF1α, HIF2α and HIF-3α) [83, 84]. Under normoxic conditions, prolyl hydroxylase domain protein 2 (PHD2) leads to hydroxylation of HIFα at two proline residues, which target via Von Hippel-Lindau (VHL) E3 ubiquitin ligase-dependent proteasomal degradation [85]. However, under the hypoxic condition, HIFα is stabilized, which is then dimerized with HIF-1β and binds HIF-response elements (HRE) to recruit gene transcription [86]. Moreover, mitochondrial DNA lacking ρ° cells are unable to stabilize HIFα proteins under hypoxic condition, which results from failure of mROS production by ETC deficiency. In contrast, MMP reconstitution restores mROS, which leads to HIFα and cell proliferation [87]. Chemical inhibition of mitochondrial ETC also attenuates mROS production in mitochondria-repleted cells, interrupting to stabilize HIFα under hypoxia [88]. Genetic loss of the complex III subunit Rieske iron-sulphur protein (RISP) or Cytc also inhibits mROS production and HIFα stabilization [89–91]. It is also indicated that mROS are requisite to activate HIFs by non-hypoxic stimulus [92].

mROS are also involved in vascular smooth muscle cell (VSMC) proliferation. Angiotensin II (AngII) is a peptide hormone basically involved in sodium and water homeostasis and vascular contraction, which is also recognized to influence cell growth and proliferation [93]. AngII exerts physiological effects by signalling via interacting with angiotensin type 1 receptors (AT1Rs) [94]. In VSMCs, AngII signalling is required to activate a multitude of mitogenic signalling cascades via crosstalk with growth factor receptors such as epidermal growth factor receptor (EGFR), platelet-derived growth factor receptors (PDGFR) and insulin receptor (IR). Intracellular signalling of VSMC proliferation is stimulated by AngII signalling-triggered mROS production and subsequently induced via mitogenic serine/threonine kinases, including ERK1/2 and p38MAPK [95].

Despite the detrimental effects, mROS function as signal transduction molecules in regulation of stem cells [96]. Depletion of ataxia telangiectasia mutated (ATM) kinase or forkhead box O (FOXO) transcriptional factors increases mROS levels, which impairs hematopoietic stem cell (HSC) proliferation [97–99]. Although the increased mROS level impairs the differentiation of HSCs, a decreased mROS level also has negative effects for self-renewal in neural and spermatogonia stem cells (SCs) [100, 101].

3.2 Differentiation

mROS function as active signalling molecules for diverse cell differentiation. Stem cells (SCs, embryonic or adult) have potentials to self-renew for maintaining stem cell pool or differentiate to the multicellular organism and supply de novo functional cells to tissues throughout the life of the organism. During differentiation of SCs, the mitochondrial oxidative metabolism is highly stimulated, and thus cellular respiration and mROS production increase [102–105].

In SCs, generally, mitochondria exhibit immature mitochondrial networks and primitive cristae [106]. As bone marrow-derived human mesenchymal stem cells (MSCs) differentiate to osteoblasts, mitochondrial biosynthesis increases by PGC-1α activation [102]. Mitochondrial mass and oxygen consumption increase during differentiation of human embryonic stem cells (ESCs) [107] or pluripotent stem cells (PSCs) [108]. Knockdown of the complex III protein RISP or mitochondrial-targeted antioxidants inhibited differentiation of human MSCs to adipocytes, indicating that mROS are required for differentiation of MSCs [109]. Furthermore, during differentiation of human PSCs, uncoupling protein 2 (UCP2) expression is repressed, which is required for metabolic transition from glycolysis to mitochondrial glucose oxidation. Knockdown of UCP2 expression facilitates mROS accumulation, which stimulate the PSC differentiation to cardiomyocytes. Ectopic UCP2 expression impairs the differentiation with retardation of mROS accumulation and embryonic body formation [110].

mROS, at least within physiological concentrations, have critical roles in processes of myogenic differentiation and muscle regeneration [111]. mROS could promote mitochondrial biogenesis, which is an essential molecule in myogenic differentiation, via peroxisome proliferator-activated receptor gamma coactivator 1 (PGC1)-activated signalling pathway [112]. Myogenic cells are armed with antioxidant enzymes such as SODs, catalase, Gpxs, Prxs, γ-glutamylcysteine synthetase (γGCS) and heme oxygenase-1 (HO-1) [113–120]. These antioxidant enzymes could play as critical signalling molecules to maintain muscle homeostasis in company with primarily neutralizing excessive ROS [121]. mROS facilitate myoblast differentiation and hypertrophy via insulin growth factor 1 (IGF1) signalling pathway [111], which enhances phosphorylation of IGF1 receptor (IGF1R) [122]. Mitochondrial complex I-derived H_2O_2 acts as a signalling molecule to induce cardiac myogenic differentiation. Chemical inhibition of the complex I and treatment of mitochondrial-specific antioxidant exhibits reduction in mROS production and thus impairs the myoblast differentiation [123]. Moreover, mROS induced phosphatase and tensin homolog (PTEN) oxidative inactivation and thereby stimulated phosphoinositide 3-kinase (PI3K)-AKT signalling pathway to express myogenic genes during skeletal myoblast differentiation and muscle regeneration [124]. In differentiation of VSMCs, mROS production also elevates to activate p38 MAPK signalling pathway [125]. However, the complexity of mROS involvement still requires further investigation to elucidate the certain roles of oxidative stress in myogenic differentiation and muscle regeneration.

3.3 Autophagy

Autophagy is a conserved catabolic process that controls cellular degradation of unnecessary or dysfunctional cellular components in the lysosome [126]. Generally, the autophagy continuously occurs to recycle damaged proteins and organelles for cellular homeostasis under normal conditions [127]. The autophagy has at least three different types: (1) Macroautophagy (usually referred to as autophagy): cytosolic contents are delivered to the lysosome by autophagosomes. (2) Microautophagy: the contents are directly introduced into lysosomal membrane. (3) Chaperone-mediated autophagy: the target proteins contain a motif KFERQ, and then the chaperone (KFERQ)-protein complex binds lysosome-associated membrane protein 2A (LAMP2A) receptors on the lysosomal membrane [128]. Autophagy induction results in recruitment of autophagy-related proteins

(ATGs) to a punctate structure, phagophore assembly site (PAS), where proteins of the uncoordinated-51-like kinase 1 (ULK1) complex assemble to initiate autophagosome formation [129].

In autophagy signalling, mitochondria are considered as main source of ROS [130]. mROS, especially as H_2O_2, are required for autophagy induction in response to nutrient starvation and rapamycin, tumour necrosis factor α (TNFα) and nerve growth factor (NGF) deprivation [131–134]. H_2O_2 modulates the cysteine protease Atg4, which cleaves c-terminus of Atg8 (or light chain 3, LC3), and thus enables the addition of phosphatidylethanolamine (PE) to Atg8. Subsequently the active Atg8 is conjugated on the autophagosomal membrane, leading to the autophagosome formation [131]. H_2O_2 also disrupts the MMP to inhibit Akt/mammalian target of rapamycin (mTOR) signalling pathway for autophagy initiation [135, 136]. Furthermore, elevated H_2O_2 induces autophagy via activation of p38 MAPK signalling pathway in cardiac or skeletal muscle [137, 138].

In physiological energy metabolism, mitochondrial ATP production by OXPHOS induces mROS generation, resulting in a certain degree of constitutive mitochondrial damage and submitochondrial particles. The damaged mitochondria cause ATP depletion and Cytc release, which eventually leads to activation of caspases and then onset of apoptosis [139, 140]. To prevent cell death, the dysfunctional mitochondria are thus sequestered from the mitochondrial network and eliminated by selective autophagy, mitophagy, to properly maintain mitochondrial quantity and quality [130]. Therefore, mitophagy limits further mROS generation, which promotes turnover of mitochondria and avoids accumulation of dysfunctional mitochondria. Mitophagy is mainly controlled by the PTEN-induced kinase 1 (PINK1)-Parkin pathway, which is stimulated upon the MMP depolarization. PINK1 is a Ser/Thr kinase that translocates on the outer mitochondrial membrane, which is stabilized by low MMP, thereby sensing mitochondrial depolarization [141–143]. Then, PINK1 recruits Parkin that ubiquitylates OMM-located proteins such as VDAC1, resulting in recruitment of autophagic machinery and the selective sequestration of ubiquitylated mitochondria within autophagosomes [130]. Furthermore, the mitochondrial proteins, BCL2/adenovirus E1B 19-kDa-interacting protein 3 (Bnip3) and Bcl-2/adenovirus E1B 19-kDa-interacting protein 3 (Bnip3L/NIX), participate in mitophagy [144]. In response to oxidative stress after ischemia/reperfusion (I/R), Bnip3 is homodimerised, to be activated, resulting in induction of mitophagy [145]. NIX, an atypical BH3 protein, is required for mitophagy in erythrocyte development. It directly recognizes autophagosome-sited GABA receptor-associated protein (GABARAP) that is a functional homolog of LC3 and subsequently induces mitophagy [126, 146]. Bnip3 and NIX directly bind to the autophagy machinery components, differently to PINK1 or Parkin [147]. ULK1 also regulates mitophagy via translocation to mitochondria to phosphorylate FUN14 domain containing 1 (FUNDC1) protein, a mitochondrial outer membrane protein, which is a receptor for hypoxia-induced mitophagy [148].

3.4 Immunity

In immune system, it is well known that ROS contribute to directly eliminate pathogens via the oxidative burst mediated by NADPH oxidases (NOXs) that are plasma membrane-bound enzyme complexes in phagosomes. However, intracellular redox state intervened by mROS has emerged to be essential for innate and adaptive immune responses [149, 150].

mROS are crucial for Toll-like receptor (TLR) signalling pathways [19]. Activation of cell surface TLRs such as TLR1, TLR2 and TLR4 increases in mROS production via TNF receptor-associated factor 6 (TRAF) and evolutionary conserved signalling intermediate in Toll pathways (ECSIT) signalling pathway [151]. The TRAF6 or ECSIT depletion promotes reduction of mROS generation in macrophages and thus impairment of bacterial clearance [151]. Lipopolysaccharide (LPS)-induced pro-inflammatory cytokines such as TNFα and IL-6 are controlled by mROS generation [152]. Innate immune response enhancement in patients with TNF receptor-associated periodic syndrome (TRAPS) that is an autoinflammatory disorder is affected by missense mutations in the type 1 TNF receptor (TNFR1), which might be attributable to mitochondrial ROS generation [152].

mROS control pattern recognition receptors (PRRs) such as nuclear oligomerization domain (NOD)-like receptors (NLRs). NLRs form multisubunit protein complexes termed inflammasomes that activate caspase-1 resulting in proteolytic cleavage and pro-inflammatory cytokine IL-1β maturation [153, 154]. Pathogen-associated molecular patterns (PAMPs) and damage-associated molecular patterns (DAMPs) such as lipopolysaccharide (LPS), asbestos, ATP and uric acid activate NLR family pyrin domain containing 3 (NLRP3) inflammasome via mROS generation [155, 156]. Pharmacologic or genetic inhibition of autophagy elevates mROS concentration, which heightens inflammasome activation [157, 158]. Increase of mROS persuades lysosomal membrane permeabilization, which is required for NLRP3 activation [159]. Activation of NLRP3 inflammasome results in mitochondrial damage, interrupting mitophagic signalling [160]. Notably, calcium influx contributes to mitochondrial damage, which might increase mROS production and mtDNA release to amplify NLRP3 inflammasome activation [161, 162]. However, it remains to be further delineated how PAMPs and DAMPs increase mROS to properly activate NLRP3 inflammasome.

In adaptive immune responses, T cells are functionally crucial in response to the pathogens [150, 156]. In infectious condition, naïve T cells promptly proliferate and differentiate into effector T cells [163]. The activation of T cells requires increase in glycolysis and mitochondrial metabolism for synthesis of macromolecules in process of the proliferation and differentiation [156, 164, 165]. Elevated mROS concentration contributes to the T-cell activation; treatment of antioxidants inhibits cellular proliferation and interleukin-2 (IL-2) production [166]. Similarly, antioxidant administration to mice exhibits their reduced immunity after infection of the virus, suggesting that mROS are indispensable for the T-cell functions in vivo [167, 168]. The T-cell receptor (TCR) stimulation induces mROS production from complex I, which leads to activation of NF-κB and AP1 signalling, and in turn facilitates IL-2 and IL-4 productions that are imperative drivers in T-cell activation [169, 170].

3.5 Aging

Aging is a process that is concomitant with the accumulation of cellular damage over the time of all living organisms. In the 1950s, Denham Harman suggested the 'free radical theory of aging' as a molecular explanation for aging [171], in which free radicals, as byproducts of energy metabolism, develop cumulative cellular damage resulting in loss of organismal ability over time. The theory has been revised that the mitochondria-derived free radicals are causative of aging [172]. Mitochondrial dysfunction and consequent excessive ROS production result in inevitable cellular damage and subsequent cell death [173]. Oxidative damage to genomes, proteins and lipids has been associated with mitochondrial dysfunction and ultimately cellular senescence or death [174]. Consistently, overexpression of antioxidant enzymes reduces ROS production and subsequently protects DNA, which is interconnected to a prolonged life span in *Drosophila melanogaster* [175, 176].

Despite numerous evidences underpinning the detrimental roles of mROS in aging, the discoveries are questioning a direct correlation between oxidative stress and the lifespan. A mitochondrial enzyme, doublecortin-like kinase 1 (MCLK1), reduction induces mitochondrial dysfunction that displays the regression of electron transport in mitochondrial respiratory chain and decline of TCA cycle activity [177]. In *Drosophila melanogaster*, mROS levels elevate along with age, but do not intervene with life span [178]. Furthermore, moderate ROS levels have been associated with an extension of longevity in *Drosophila melanogaster* and in young mice [179–181]. Therefore, physiologically controlled mROS might activate adaptive responses that are beneficial to the organism and extend life span.

4. Conclusion

Mitochondria are primary energy producers to generate ATPs via oxidative phosphorylation. For a long time, mROS have been considered as byproducts of biological energy metabolism during the ATP generation or by cellular redox system imbalance, which are highly aggressive and detrimental to the neighbouring cells and tissues. However, the roles of mROS have been extensively substantiated to understand normal physiology and pathology over the past decades. Mitochondria-derived H_2O_2 have been unequivocally recognized as essential molecules in a range of physiological processes in cells.

In this chapter, we have provided a brief discussion of current understanding of physiological roles of mROS by which mitochondria indeed contribute to the implementation of cellular proliferation, differentiation autophagy, innate and adaptive immunity and aging. In understanding the mechanisms regulating mitochondrial physiology and homeostasis, mROS production might provide a significant potential for the development of novel therapeutic strategy for the treatment of a wide range of human pathologies.

Acknowledgements

This work was supported by the National Research Foundation of Korea (NRF) grant funded by the Korean government (MEST) (NRF-2013R1A1A2061214, NRF-2018R1D1A1B07048909 and NRF-2018R1A6A1A03025124).

Conflict of interest

The authors declare no conflict of interest.

Author details

Tae Gyu Choi* and Sung Soo Kim
Department of Biochemistry and Molecular Biology, School of Medicine,
Kyung Hee University, Seoul, Republic of Korea

*Address all correspondence to: aske@daum.net

References

[1] Henze K, Martin W. Evolutionary biology: Essence of mitochondria. Nature. 2003;**426**:127-128

[2] Zorov DB, Juhaszova M, Sollott SJ. Mitochondrial reactive oxygen species (ROS) and ROS-induced ROS release. Physiological Reviews. 2014;**94**:909-950

[3] Brand MD, Nicholls DG. Assessing mitochondrial dysfunction in cells. The Biochemical Journal. 2011;**435**:297-312

[4] Balaban RS, Nemoto S, Finkel T. Mitochondria, oxidants, and aging. Cell. 2005;**120**:483-495

[5] Brand MD. Mitochondrial generation of superoxide and hydrogen peroxide as the source of mitochondrial redox signaling. Free Radical Biology & Medicine. 2016;**100**:14-31

[6] Thannickal VJ, Fanburg BL. Reactive oxygen species in cell signaling. American Journal of Physiology. Lung Cellular and Molecular Physiology. 2000;**279**:L1005-L1028

[7] Li X, Fang P, Mai J, Choi ET, Wang H, Yang XF. Targeting mitochondrial reactive oxygen species as novel therapy for inflammatory diseases and cancers. Journal of Hematology & Oncology. 2013;**6**:19

[8] Hayyan M, Hashim MA, AlNashef IM. Superoxide ion: Generation and chemical implications. Chemical Reviews. 2016;**116**:3029-3085

[9] Murphy MP. How mitochondria produce reactive oxygen species. The Biochemical Journal. 2009;**417**: 1-13

[10] Finkel T. Signal transduction by reactive oxygen species. The Journal of Cell Biology. 2011;**194**:7-15

[11] Chandel NS. Mitochondrial regulation of oxygen sensing. Advances in Experimental Medicine and Biology. 2010;**661**:339-354

[12] Antico Arciuch VG, Elguero ME, Poderoso JJ, Carreras MC. Mitochondrial regulation of cell cycle and proliferation. Antioxidants & Redox Signaling. 2012;**16**:1150-1180

[13] Rambold AS, Pearce EL. Mitochondrial dynamics at the Interface of immune cell metabolism and function. Trends in Immunology. 2018;**39**:6-18

[14] Gunter TE, Yule DI, Gunter KK, Eliseev RA, Salter JD. Calcium and mitochondria. FEBS Letters. 2004;**567**:96-102

[15] Nikoletopoulou V, Markaki M, Palikaras K, Tavernarakis N. Crosstalk between apoptosis, necrosis and autophagy. Biochimica et Biophysica Acta. 2013;**1833**:3448-3459

[16] Kamer KJ, Mootha VK. The molecular era of the mitochondrial calcium uniporter. Nature Reviews. Molecular Cell Biology. 2015;**16**:545-553

[17] Droge W. Free radicals in the physiological control of cell function. Physiological Reviews. 2002;**82**:47-95

[18] Finkel T. Oxygen radicals and signaling. Current Opinion in Cell Biology. 1998;**10**:248-253

[19] Sena LA, Chandel NS. Physiological roles of mitochondrial reactive oxygen species. Molecular Cell. 2012;**48**:158-167

[20] Schieber M, Chandel NS. ROS function in redox signaling and oxidative stress. Current Biology. 2014;**24**:R453-R462

[21] McCommis KS, Finck BN. Mitochondrial pyruvate transport: A historical perspective and future research directions. The Biochemical Journal. 2015;**466**:443-454

[22] Friedman JR, Nunnari J. Mitochondrial form and function. Nature. 2014;**505**:335-343

[23] Lane N, Martin W. The energetics of genome complexity. Nature. 2010;**467**:929-934

[24] Burki F. Mitochondrial evolution: Going, going, gone. Current Biology. 2016;**26**:R410-R412

[25] Gabaldon T, Huynen MA. Shaping the mitochondrial proteome. Biochimica et Biophysica Acta. 2004;**1659**:212-220

[26] Taanman JW. The mitochondrial genome: Structure, transcription, translation and replication. Biochimica et Biophysica Acta. 1999;**1410**:103-123

[27] Youle RJ, van der Bliek AM. Mitochondrial fission, fusion, and stress. Science. 2012;**337**:1062-1065

[28] Lackner LL. Shaping the dynamic mitochondrial network. BMC Biology. 2014;**12**:35

[29] Pernas L, Scorrano L. Mito-morphosis: Mitochondrial fusion fission, and cristae remodeling as key mediators of cellular function. Annual Review of Physiology. 2016; **78**:505-531

[30] Jap BK, Walian PJ. Structure and functional mechanism of porins. Physiological Reviews. 1996;**76**:1073-1088

[31] Hoogenboom BW, Suda K, Engel A, Fotiadis D. The supramolecular assemblies of voltage-dependent anion channels in the native membrane. Journal of Molecular Biology. 2007;**370**:246-255

[32] Zeth K. Structure and evolution of mitochondrial outer membrane proteins of beta-barrel topology. Biochimica et Biophysica Acta. 2010;**1797**:1292-1299

[33] Mannella CA. Structure and dynamics of the mitochondrial inner membrane cristae. Biochimica et Biophysica Acta. 2006;**1763**:542-548

[34] Wohlrab H. Transport proteins (carriers) of mitochondria. IUBMB Life. 2009;**61**:40-46

[35] Herrmann JM, Riemer J. The intermembrane space of mitochondria. Antioxidants & Redox Signaling. 2010;**13**:1341-1358

[36] O'Rourke B. Mitochondrial ion channels. Annual Review of Physiology. 2007;**69**:19-49

[37] Backes S, Herrmann JM. Protein translocation into the intermembrane space and matrix of mitochondria: Mechanisms and driving forces. Frontiers in Molecular Biosciences. 2017;**4**:83

[38] Picard M, Taivassalo T, Gouspillou G, Hepple RT. Mitochondria: Isolation, structure and function. The Journal of Physiology. 2011;**589**: 4413-4421

[39] Boore JL. Animal mitochondrial genomes. Nucleic Acids Research. 1999;**27**:1767-1780

[40] Blanco FJ, Valdes AM, Rego-Perez I. Mitochondrial DNA variation and the pathogenesis of osteoarthritis phenotypes. Nature Reviews Rheumatology. 2018;**14**:327-340

[41] Anderson S, Bankier AT, Barrell BG, de Bruijn MH, Coulson AR, Drouin J, et al. Sequence and organization of the human mitochondrial genome. Nature. 1981;**290**:457-465

[42] Wallace DC. Maternal genes: Mitochondrial diseases. Birth Defects Original Article Series. 1987;**23**:137-190

[43] Dautant A, Meier T, Hahn A, Tribouillard-Tanvier D, di Rago JP, Kucharczyk R. ATP synthase diseases of mitochondrial genetic origin. Frontiers in Physiology. 2018;**9**:329

[44] Park CB, Larsson NG. Mitochondrial DNA mutations in disease and aging. The Journal of Cell Biology. 2011;**193**:809-818

[45] Madamanchi NR, Runge MS. Mitochondrial dysfunction in atherosclerosis. Circulation Research. 2007;**100**:460-473

[46] Terzioglu M, Larsson NG. Mitochondrial dysfunction in mammalian ageing. Novartis Foundation Symposium. 2007;**287**: 197-208; discussion 208-113

[47] Liesa M, Shirihai OS. Mitochondrial dynamics in the regulation of nutrient utilization and energy expenditure. Cell Metabolism. 2013;**17**:491-506

[48] Herrera AS, Del CAEM, Md Ashraf G, Zamyatnin AA, Aliev G. Beyond mitochondria, what would be the energy source of the cell? Central Nervous System Agents in Medicinal Chemistry. 2015;**15**:32-41

[49] Anderson NM, Mucka P, Kern JG, Feng H. The emerging role and targetability of the TCA cycle in cancer metabolism. Protein & Cell. 2018;**9**:216-237

[50] Cogliati S, Enriquez JA, Scorrano L. Mitochondrial cristae: Where beauty meets functionality. Trends in Biochemical Sciences. 2016;**41**:261-273

[51] van der Bliek AM, Shen Q, Kawajiri S. Mechanisms of mitochondrial fission and fusion. Cold Spring Harbor Perspectives in Biology. 2013;**5**:30-33

[52] Bartolak-Suki E, Imsirovic J, Nishibori Y, Krishnan R, Suki B. Regulation of mitochondrial structure and dynamics by the cytoskeleton and mechanical factors. International Journal of Molecular Sciences. 2017;**18**:34-36

[53] Yaffe MP. The machinery of mitochondrial inheritance and behavior. Science. 1999;**283**:1493-1497

[54] Bach D, Pich S, Soriano FX, Vega N, Baumgartner B, Oriola J, et al. Mitofusin-2 determines mitochondrial network architecture and mitochondrial metabolism. A novel regulatory mechanism altered in obesity. The Journal of Biological Chemistry. 2003;**278**:17190-17197

[55] Scott I, Youle RJ. Mitochondrial fission and fusion. Essays in Biochemistry. 2010;**47**:85-98

[56] Santel A, Fuller MT. Control of mitochondrial morphology by a human mitofusin. Journal of Cell Science. 2001;**114**:867-874

[57] Chan DC. Fusion and fission: Interlinked processes critical for mitochondrial health. Annual Review of Genetics. 2012;**46**:265-287

[58] Westermann B. Mitochondrial fusion and fission in cell life and death. Nature Reviews. Molecular Cell Biology. 2010;**11**:872-884

[59] Cadenas E, Davies KJ. Mitochondrial free radical generation, oxidative stress, and aging. Free Radical Biology & Medicine. 2000;**29**:222-230

[60] Kausar S, Wang F, Cui H. The role of mitochondria in reactive oxygen species generation and its implications for neurodegenerative diseases. Cells. 2018;**7**:5-7

[61] Starkov AA. The role of mitochondria in reactive oxygen species metabolism and signaling. Annals of the New York Academy of Sciences. 2008;**1147**:37-52

[62] Kalogeris T, Bao Y, Korthuis RJ. Mitochondrial reactive oxygen species: A double edged sword in ischemia/ reperfusion vs preconditioning. Redox Biology. 2014;**2**:702-714

[63] Cox AG, Winterbourn CC, Hampton MB. Mitochondrial peroxire-doxin involvement in antioxidant defence and redox signalling. The Biochemical Journal. 2009;**425**:313-325

[64] Andreyev AY, Kushnareva YE, Murphy AN, Starkov AA. Mitochondrial ROS metabolism: 10 years later. Biochemistry (Mosc). 2015;**80**:517-531

[65] Liemburg-Apers DC, Willems PH, Koopman WJ, Grefte S. Interactions between mitochondrial reactive oxygen species and cellular glucose metabolism. Archives of Toxicology. 2015;**89**:1209-1226

[66] Birben E, Sahiner UM, Sackesen C, Erzurum S, Kalayci O. Oxidative stress and antioxidant defense. World Allergy Organization Journal. 2012;**5**:9-19

[67] Kirkman HN, Rolfo M, Ferraris AM, Gaetani GF. Mechanisms of protection of catalase by NADPH. Kinetics and stoichiometry. The Journal of Biological Chemistry. 1999;**274**:13908-13914

[68] Stairs CW, Leger MM, Roger AJ. Diversity and origins of anaerobic metabolism in mitochondria and related organelles. Philosophical Transactions of the Royal Society of London. Series B, Biological Sciences. 2015;**370**:20140326

[69] Angelova PR, Abramov AY. Role of mitochondrial ROS in the brain: From physiology to neurodegeneration. FEBS Letters. 2018;**592**:692-702

[70] Kornfeld OS, Hwang S, Disatnik MH, Chen CH, Qvit N, Mochly-Rosen D. Mitochondrial reactive oxygen species at the heart of the matter: New therapeutic approaches for cardiovascular diseases. Circulation Research. 2015;**116**:1783-1799

[71] Nazarewicz RR, Dikalova A, Bikineyeva A, Ivanov S, Kirilyuk IA, Grigor'ev IA, et al. Does scavenging of mitochondrial superoxide attenuate cancer prosurvival signaling pathways? Antioxidants & Redox Signaling. 2013;**19**:344-349

[72] Zorov DB, Bannikova SY, Belousov VV, Vyssokikh MY, Zorova LD, Isaev NK, et al. Reactive oxygen and nitrogen species: Friends or foes? Biochemistry (Mosc). 2005;**70**: 215-221

[73] Runkel ED, Baumeister R, Schulze E. Mitochondrial stress: Balancing friend and foe. Experimental Gerontology. 2014;**56**:194-201

[74] Janssen-Heininger YM, Mossman BT, Heintz NH, Forman HJ, Kalyanaraman B, Finkel T, et al. Redox-based regulation of signal transduction: Principles, pitfalls, and promises. Free Radical Biology & Medicine. 2008;**45**:1-17

[75] D'Autreaux B, Toledano MB. ROS as signalling molecules: Mechanisms that generate specificity in ROS homeostasis. Nature Reviews. Molecular Cell Biology. 2007;**8**:813-824

[76] DieboldL,ChandelNS.Mitochondrial ROS regulation of proliferating cells. Free Radical Biology & Medicine. 2016;**100**:86-93

[77] Brandes N, Schmitt S, Jakob U. Thiol-based redox switches in

eukaryotic proteins. Antioxidants & Redox Signaling. 2009;**11**:997-1014

[78] Finkel T. From sulfenylation to sulfhydration: What a thiolate needs to tolerate. Science Signaling. 2012;**5**:pe10

[79] Nagiec EE, Wu L, Swaney SM, Chosay JG, Ross DE, Brieland JK, et al. Oxazolidinones inhibit cellular proliferation via inhibition of mitochondrial protein synthesis. Antimicrobial Agents and Chemotherapy. 2005;**49**:3896-3902

[80] Onyango IG, Khan SM, Bennett JP Jr. Mitochondria in the pathophysiology of Alzheimer's and Parkinson's diseases. Frontiers in Bioscience (Landmark Ed). 2017;**22**:854-872

[81] Haase VH. Hypoxic regulation of erythropoiesis and iron metabolism. American Journal of Physiology. Renal Physiology. 2010;**299**:F1-F13

[82] Beleslin-Cokic BB, Cokic VP, Yu X, Weksler BB, Schechter AN, Noguchi CT. Erythropoietin and hypoxia stimulate erythropoietin receptor and nitric oxide production by endothelial cells. Blood. 2004;**104**:2073-2080

[83] Weidemann A, Johnson RS. Biology of HIF-1alpha. Cell Death and Differentiation. 2008;**15**:621-627

[84] Kaelin WG Jr, Ratcliffe PJ. Oxygen sensing by metazoans: The central role of the HIF hydroxylase pathway. Molecular Cell. 2008;**30**:393-402

[85] Semenza GL. Hypoxia-inducible factor 1 (HIF-1) pathway. Science's STKE. 2007;**2007**:cm8

[86] Semenza GL. Hypoxia-inducible factors in physiology and medicine. Cell. 2012;**148**:399-408

[87] Martinez-Reyes I, Diebold LP, Kong H, Schieber M, Huang H, Hensley CT, et al. TCA cycle and mitochondrial membrane potential are necessary for diverse biological functions. Molecular Cell. 2016;**61**:199-209

[88] Chandel NS, McClintock DS, Feliciano CE, Wood TM, Melendez JA, Rodriguez AM, et al. Reactive oxygen species generated at mitochondrial complex III stabilize hypoxia-inducible factor-1alpha during hypoxia: A mechanism of O_2 sensing. The Journal of Biological Chemistry. 2000;**275**:25130-25138

[89] Brunelle JK, Bell EL, Quesada NM, Vercauteren K, Tiranti V, Zeviani M, et al. Oxygen sensing requires mitochondrial ROS but not oxidative phosphorylation. Cell Metabolism. 2005;**1**:409-414

[90] Guzy RD, Hoyos B, Robin E, Chen H, Liu L, Mansfield KD, et al. Mitochondrial complex III is required for hypoxia-induced ROS production and cellular oxygen sensing. Cell Metabolism. 2005;**1**:401-408

[91] Mansfield KD, Guzy RD, Pan Y, Young RM, Cash TP, Schumacker PT, et al. Mitochondrial dysfunction resulting from loss of cytochrome c impairs cellular oxygen sensing and hypoxic HIF-alpha activation. Cell Metabolism. 2005;**1**:393-399

[92] Patten DA, Lafleur VN, Robitaille GA, Chan DA, Giaccia AJ, Richard DE. Hypoxia-inducible factor-1 activation in nonhypoxic conditions: The essential role of mitochondrial-derived reactive oxygen species. Molecular Biology of the Cell. 2010;**21**:3247-3257

[93] Wolf G, Wenzel UO. Angiotensin II and cell cycle regulation. Hypertension. 2004;**43**:693-698

[94] Kim S, Zingler M, Harrison JK, Scott EW, Cogle CR, Luo D, et al.

Angiotensin II regulation of proliferation, differentiation, and engraftment of hematopoietic stem cells. Hypertension. 2016;**67**:574-584

[95] Mehta PK, Griendling KK. Angiotensin II cell signaling: Physiological and pathological effects in the cardiovascular system. American Journal of Physiology. Cell Physiology. 2007;**292**:C82-C97

[96] Holmstrom KM, Finkel T. Cellular mechanisms and physiological consequences of redox-dependent signalling. Nature Reviews. Molecular Cell Biology. 2014;**15**:411-421

[97] Ito K, Hirao A, Arai F, Matsuoka S, Takubo K, Hamaguchi I, et al. Regulation of oxidative stress by ATM is required for self-renewal of haematopoietic stem cells. Nature. 2004;**431**:997-1002

[98] Miyamoto K, Araki KY, Naka K, Arai F, Takubo K, Yamazaki S, et al. Foxo3a is essential for maintenance of the hematopoietic stem cell pool. Cell Stem Cell. 2007;**1**:101-112

[99] Tothova Z, Kollipara R, Huntly BJ, Lee BH, Castrillon DH, Cullen DE, et al. FoxOs are critical mediators of hematopoietic stem cell resistance to physiologic oxidative stress. Cell. 2007;**128**:325-339

[100] Morimoto H, Iwata K, Ogonuki N, Inoue K, Atsuo O, Kanatsu-Shinohara M, et al. ROS are required for mouse spermatogonial stem cell self-renewal. Cell Stem Cell. 2013;**12**:774-786

[101] Le Belle JE, Orozco NM, Paucar AA, Saxe JP, Mottahedeh J, Pyle AD, et al. Proliferative neural stem cells have high endogenous ROS levels that regulate self-renewal and neurogenesis in a PI3K/Akt-dependant manner. Cell Stem Cell. 2011;**8**:59-71

[102] Chen CT, Shih YR, Kuo TK, Lee OK, Wei YH. Coordinated changes of mitochondrial biogenesis and antioxidant enzymes during osteogenic differentiation of human mesenchymal stem cells. Stem Cells. 2008;**26**:960-968

[103] Chung S, Dzeja PP, Faustino RS, Perez-Terzic C, Behfar A, Terzic A. Mitochondrial oxidative metabolism is required for the cardiac differentiation of stem cells. Nature Clinical Practice. Cardiovascular Medicine. 2007;**4**(Suppl 1):S60-S67

[104] Lonergan T, Brenner C, Bavister B. Differentiation-related changes in mitochondrial properties as indicators of stem cell competence. Journal of Cellular Physiology. 2006;**208**:149-153

[105] Khacho M, Clark A, Svoboda DS, Azzi J, MacLaurin JG, Meghaizel C, et al. Mitochondrial dynamics impacts stem cell identity and fate decisions by regulating a nuclear transcriptional program. Cell Stem Cell. 2016;**19**:232-247

[106] Papa L, Djedaini M, Hoffman R. Mitochondrial role in stemness and differentiation of hematopoietic stem cells. Stem Cells International. 2019;**2019**:4067162

[107] Cho YM, Kwon S, Pak YK, Seol HW, Choi YM, Park DJ, et al. Dynamic changes in mitochondrial biogenesis and antioxidant enzymes during the spontaneous differentiation of human embryonic stem cells. Biochemical and Biophysical Research Communications. 2006;**348**:1472-1478

[108] Prigione A, Adjaye J. Modulation of mitochondrial biogenesis and bioenergetic metabolism upon in vitro and in vivo differentiation of human ES and iPS cells. The International Journal of Developmental Biology. 2010;**54**:1729-1741

[109] Tormos KV, Anso E, Hamanaka RB, Eisenbart J, Joseph J, Kalyanaraman B, et al. Mitochondrial complex III ROS regulate adipocyte differentiation. Cell Metabolism. 2011;**14**:537-544

[110] Zhang J, Khvorostov I, Hong JS, Oktay Y, Vergnes L, Nuebel E, et al. UCP2 regulates energy metabolism and differentiation potential of human pluripotent stem cells. The EMBO Journal. 2011;**30**:4860-4873

[111] Barbieri E, Sestili P. Reactive oxygen species in skeletal muscle signaling. Journal of Signal Transduction. 2012;**2012**:982794

[112] Adhihetty PJ, Irrcher I, Joseph AM, Ljubicic V, Hood DA. Plasticity of skeletal muscle mitochondria in response to contractile activity. Experimental Physiology. 2003;**88**:99-107

[113] Powers SK, Criswell D, Lawler J, Ji LL, Martin D, Herb RA, et al. Influence of exercise and fiber type on antioxidant enzyme activity in rat skeletal muscle. The American Journal of Physiology. 1994;**266**:R375-R380

[114] Manabe Y, Takagi M, Nakamura-Yamada M, Goto-Inoue N, Taoka M, Isobe T, et al. Redox proteins are constitutively secreted by skeletal muscle. The Journal of Physiological Sciences. 2014;**64**:401-409

[115] Tidball JG. Mechanisms of muscle injury, repair, and regeneration. Comprehensive Physiology. 2011;**1**:2029-2062

[116] El Haddad M, Jean E, Turki A, Hugon G, Vernus B, Bonnieu A, et al. Glutathione peroxidase 3, a new retinoid target gene, is crucial for human skeletal muscle precursor cell survival. Journal of Cell Science. 2012;**125**:6147-6156

[117] Kozakowska M, Ciesla M, Stefanska A, Skrzypek K, Was H, Jazwa A, et al. Heme oxygenase-1 inhibits myoblast differentiation by targeting myomirs. Antioxidants & Redox Signaling. 2012;**16**:113-127

[118] Ding Y, Choi KJ, Kim JH, Han X, Piao Y, Jeong JH, et al. Endogenous hydrogen peroxide regulates glutathione redox via nuclear factor erythroid 2-related factor 2 downstream of phosphatidylinositol 3-kinase during muscle differentiation. The American Journal of Pathology. 2008;**172**:1529-1541

[119] Catani MV, Savini I, Duranti G, Caporossi D, Ceci R, Sabatini S, et al. Nuclear factor kappaB and activating protein 1 are involved in differentiation-related resistance to oxidative stress in skeletal muscle cells. Free Radical Biology & Medicine. 2004;**37**:1024-1036

[120] Won H, Lim S, Jang M, Kim Y, Rashid MA, Jyothi KR, et al. Peroxiredoxin-2 upregulated by NF-kappaB attenuates oxidative stress during the differentiation of muscle-derived C2C12 cells. Antioxidants & Redox Signaling. 2012;**16**:245-261

[121] Kozakowska M, Pietraszek-Gremplewicz K, Jozkowicz A, Dulak J. The role of oxidative stress in skeletal muscle injury and regeneration: Focus on antioxidant enzymes. Journal of Muscle Research and Cell Motility. 2015;**36**:377-393

[122] Handayaningsih A-E, et al. Reactive Oxygen Species Play an Essential Role in IGF-I Signaling and IGF-I-Induced Myocyte Hypertrophy in C2C12 Myocytes. Endocrinology. 2011;**152**(3). DOI: 10.1210/en.2010-0981

[123] Lee S, Tak E, Lee J, Rashid MA, Murphy MP, Ha J, et al. Mitochondrial H_2O_2 generated from electron transport chain complex I stimulates muscle differentiation. Cell Research. 2011;**21**:817-834

[124] Kim JH, Choi TG, Park S, Yun HR, Nguyen NNY, Jo YH, et al. Mitochondrial ROS-derived PTEN oxidation activates PI3K pathway for mTOR-induced myogenic autophagy. Cell Death and Differentiation. 2018;**25**:1921-1937

[125] Clempus RE, Griendling KK. Reactive oxygen species signaling in vascular smooth muscle cells. Cardiovascular Research. 2006;**71**:216-225

[126] Filomeni G, De Zio D, Cecconi F. Oxidative stress and autophagy: The clash between damage and metabolic needs. Cell Death and Differentiation. 2015;**22**:377-388

[127] Lee J, Giordano S, Zhang J. Autophagy, mitochondria and oxidative stress: Cross-talk and redox signalling. The Biochemical Journal. 2012;**441**:523-540

[128] Parzych KR, Klionsky DJ. An overview of autophagy: Morphology, mechanism, and regulation. Antioxidants & Redox Signaling. 2014;**20**:460-473

[129] Hurley JH, Young LN. Mechanisms of autophagy initiation. Annual Review of Biochemistry. 2017;**86**:225-244

[130] Dikic I, Elazar Z. Mechanism and medical implications of mammalian autophagy. Nature Reviews. Molecular Cell Biology. 2018;**19**:349-364

[131] Scherz-Shouval R, Shvets E, Fass E, Shorer H, Gil L, Elazar Z. Reactive oxygen species are essential for autophagy and specifically regulate the activity of Atg4. The EMBO Journal. 2007;**26**:1749-1760

[132] Djavaheri-Mergny M, Amelotti M, Mathieu J, Besancon F, Bauvy C, Souquere S, et al. NF-kappaB activation represses tumor necrosis factor-alpha-induced autophagy. The Journal of Biological Chemistry. 2006;**281**:30373-30382

[133] Kirkland RA, Saavedra GM, Franklin JL. Rapid activation of antioxidant defenses by nerve growth factor suppresses reactive oxygen species during neuronal apoptosis: Evidence for a role in cytochrome c redistribution. The Journal of Neuroscience. 2007;**27**:11315-11326

[134] Kirkland RA, Adibhatla RM, Hatcher JF, Franklin JL. Loss of cardiolipin and mitochondria during programmed neuronal death: Evidence of a role for lipid peroxidation and autophagy. Neuroscience. 2002;**115**:587-602

[135] Byun YJ, Kim SK, Kim YM, Chae GT, Jeong SW, Lee SB. Hydrogen peroxide induces autophagic cell death in C6 glioma cells via BNIP3-mediated suppression of the mTOR pathway. Neuroscience Letters. 2009;**461**:131-135

[136] Zhang H, Kong X, Kang J, Su J, Li Y, Zhong J, et al. Oxidative stress induces parallel autophagy and mitochondria dysfunction in human glioma U251 cells. Toxicological Sciences. 2009;**110**:376-388

[137] Yuan H, Perry CN, Huang C, Iwai-Kanai E, Carreira RS, Glembotski CC, et al. LPS-induced autophagy is mediated by oxidative signaling in cardiomyocytes and is associated with cytoprotection. American Journal of Physiology. Heart and Circulatory Physiology. 2009;**296**:H470-H479

[138] McClung JM, Judge AR, Powers SK, Yan Z. p38 MAPK links oxidative stress to autophagy-related gene expression in cachectic muscle wasting. American Journal of Physiology. Cell Physiology. 2010;**298**:C542-C549

[139] Mao K, Klionsky DJ. Participation of mitochondrial fission during

mitophagy. Cell Cycle. 2013;**12**:3131-3132

[140] Bolisetty S, Jaimes EA. Mitochondria and reactive oxygen species: Physiology and pathophysiology. International Journal of Molecular Sciences. 2013;**14**:6306-6344

[141] Ordureau A, Sarraf SA, Duda DM, Heo JM, Jedrychowski MP, Sviderskiy VO, et al. Quantitative proteomics reveal a feedforward mechanism for mitochondrial PARKIN translocation and ubiquitin chain synthesis. Molecular Cell. 2014;**56**:360-375

[142] Shiba-Fukushima K, Arano T, Matsumoto G, Inoshita T, Yoshida S, Ishihama Y, et al. Phosphorylation of mitochondrial polyubiquitin by PINK1 promotes Parkin mitochondrial tethering. PLoS Genetics. 2014;**10**:e1004861

[143] Lazarou M, Sliter DA, Kane LA, Sarraf SA, Wang C, Burman JL, et al. The ubiquitin kinase PINK1 recruits autophagy receptors to induce mitophagy. Nature. 2015;**524**:309-314

[144] Ney PA. Mitochondrial autophagy: Origins, significance, and role of BNIP3 and NIX. Biochimica et Biophysica Acta. 2015;**1853**:2775-2783

[145] Saito T, Sadoshima J. Molecular mechanisms of mitochondrial autophagy/mitophagy in the heart. Circulation Research. 2015;**116**:1477-1490

[146] Novak I, Kirkin V, McEwan DG, Zhang J, Wild P, Rozenknop A, et al. Nix is a selective autophagy receptor for mitochondrial clearance. EMBO Reports. 2010;**11**:45-51

[147] Matsuda N. Phospho-ubiquitin: Upending the PINK-Parkin-ubiquitin cascade. Journal of Biochemistry. 2016;**159**:379-385

[148] Wu W, Tian W, Hu Z, Chen G, Huang L, Li W, et al. ULK1 translocates to mitochondria and phosphorylates FUNDC1 to regulate mitophagy. EMBO Reports. 2014;**15**:566-575

[149] Chen Y, Zhou Z, Min W. Mitochondria, oxidative stress and innate immunity. Frontiers in Physiology. 2018;**9**:1487

[150] Yang Y, Bazhin AV, Werner J, Karakhanova S. Reactive oxygen species in the immune system. International Reviews of Immunology. 2013;**32**:249-270

[151] West AP, Brodsky IE, Rahner C, Woo DK, Erdjument-Bromage H, Tempst P, et al. TLR signalling augments macrophage bactericidal activity through mitochondrial ROS. Nature. 2011;**472**:476-480

[152] Bulua AC, Simon A, Maddipati R, Pelletier M, Park H, Kim KY, et al. Mitochondrial reactive oxygen species promote production of proinflammatory cytokines and are elevated in TNFR1-associated periodic syndrome (TRAPS). The Journal of Experimental Medicine. 2011;**208**:519-533

[153] Schroder K, Tschopp J. The inflammasomes. Cell. 2010;**140**:821-832

[154] Broz P, Dixit VM. Inflammasomes: Mechanism of assembly, regulation and signalling. Nature Reviews. Immunology. 2016;**16**:407-420

[155] Li F, Xu M, Wang M, Wang L, Wang H, Zhang H, et al. Roles of mitochondrial ROS and NLRP3 inflammasome in multiple ozone-induced lung inflammation and emphysema. Respiratory Research. 2018;**19**:230

[156] Weinberg SE, Sena LA, Chandel NS. Mitochondria in the

regulation of innate and adaptive immunity. Immunity. 2015;**42**:406-417

[157] Saitoh T, Fujita N, Jang MH, Uematsu S, Yang BG, Satoh T, et al. Loss of the autophagy protein Atg16L1 enhances endotoxin-induced IL-1beta production. Nature. 2008;**456**:264-268

[158] Liu Q, Zhang D, Hu D, Zhou X, Zhou Y. The role of mitochondria in NLRP3 inflammasome activation. Molecular Immunology. 2018;**103**:115-124

[159] Heid ME, Keyel PA, Kamga C, Shiva S, Watkins SC, Salter RD. Mitochondrial reactive oxygen species induces NLRP3-dependent lysosomal damage and inflammasome activation. Journal of Immunology. 2013;**191**:5230-5238

[160] Yu J, Nagasu H, Murakami T, Hoang H, Broderick L, Hoffman HM, et al. Inflammasome activation leads to Caspase-1-dependent mitochondrial damage and block of mitophagy. Proceedings of the National Academy of Sciences of the United States of America. 2014;**111**:15514-15519

[161] Gurung P, Lukens JR, Kanneganti TD. Mitochondria: Diversity in the regulation of the NLRP3 inflammasome. Trends in Molecular Medicine. 2015;**21**:193-201

[162] Murakami T, Ockinger J, Yu J, Byles V, McColl A, Hofer AM, et al. Critical role for calcium mobilization in activation of the NLRP3 inflammasome. Proceedings of the National Academy of Sciences of the United States of America. 2012;**109**:11282-11287

[163] Smith-Garvin JE, Koretzky GA, Jordan MS. T cell activation. Annual Review of Immunology. 2009;**27**:591-619

[164] Carr EL, Kelman A, Wu GS, Gopaul R, Senkevitch E, Aghvanyan A, et al. Glutamine uptake and metabolism are coordinately regulated by ERK/MAPK during T lymphocyte activation. Journal of Immunology. 2010;**185**:1037-1044

[165] Sinclair LV, Rolf J, Emslie E, Shi YB, Taylor PM, Cantrell DA. Control of amino-acid transport by antigen receptors coordinates the metabolic reprogramming essential for T cell differentiation. Nature Immunology. 2013;**14**:500-508

[166] Chaudhri G, Clark IA, Hunt NH, Cowden WB, Ceredig R. Effect of antioxidants on primary alloantigen-induced T cell activation and proliferation. Journal of Immunology. 1986;**137**:2646-2652

[167] Laniewski NG, Grayson JM. Antioxidant treatment reduces expansion and contraction of antigen-specific CD8+ T cells during primary but not secondary viral infection. Journal of Virology. 2004;**78**:11246-11257

[168] Sena LA, Li S, Jairaman A, Prakriya M, Ezponda T, Hildeman DA, et al. Mitochondria are required for antigen-specific T cell activation through reactive oxygen species signaling. Immunity. 2013;**38**:225-236

[169] Gill T, Levine AD. Mitochondria-derived hydrogen peroxide selectively enhances T cell receptor-initiated signal transduction. The Journal of Biological Chemistry. 2013;**288**:26246-26255

[170] Kaminski MM, Sauer SW, Klemke CD, Suss D, Okun JG, Krammer PH, et al. Mitochondrial reactive oxygen species control T cell activation by regulating IL-2 and IL-4 expression: Mechanism of ciprofloxacin-mediated immunosuppression. Journal of Immunology. 2010;**184**:4827-4841

[171] Harman D. Aging: A theory based on free radical and radiation

chemistry. Journal of Gerontology. 1956;**11**:298-300

[172] Harman D. The biologic clock: The mitochondria? Journal of the American Geriatrics Society. 1972;**20**:145-147

[173] Ziegler DV, Wiley CD, Velarde MC. Mitochondrial effectors of cellular senescence: Beyond the free radical theory of aging. Aging Cell. 2015;**14**:1-7

[174] Bokov A, Chaudhuri A, Richardson A. The role of oxidative damage and stress in aging. Mechanisms of Ageing and Development. 2004;**125**:811-826

[175] Orr WC, Sohal RS. Extension of life-span by overexpression of superoxide dismutase and catalase in Drosophila melanogaster. Science. 1994;**263**:1128-1130

[176] Schriner SE, Linford NJ, Martin GM, Treuting P, Ogburn CE, Emond M, et al. Extension of murine life span by overexpression of catalase targeted to mitochondria. Science. 2005;**308**:1909-1911

[177] Lapointe J, Hekimi S. Early mitochondrial dysfunction in long-lived Mclk1$^{+/-}$ mice. The Journal of Biological Chemistry. 2008;**283**:26217-26227

[178] Cocheme HM, Quin C, McQuaker SJ, Cabreiro F, Logan A, Prime TA, et al. Measurement of H_2O_2 within living Drosophila during aging using a ratiometric mass spectrometry probe targeted to the mitochondrial matrix. Cell Metabolism. 2011;**13**:340-350

[179] Copeland JM, Cho J, Lo T Jr, Hur JH, Bahadorani S, Arabyan T, et al. Extension of Drosophila life span by RNAi of the mitochondrial respiratory chain. Current Biology. 2009;**19**: 1591-1598

[180] Basisty N, Dai DF, Gagnidze A, Gitari L, Fredrickson J, Maina Y, et al. Mitochondrial-targeted catalase is good for the old mouse proteome, but not for the young: 'Reverse' antagonistic pleiotropy? Aging Cell. 2016;**15**:634-645

[181] Csiszar A, Labinskyy N, Perez V, Recchia FA, Podlutsky A, Mukhopadhyay P, et al. Endothelial function and vascular oxidative stress in long-lived GH/IGF-deficient Ames dwarf mice. American Journal of Physiology. Heart and Circulatory Physiology. 2008;**295**:H1882-H1894

Chapter 3

Role of Oxidative Stress in the Cardiovascular Complications of Kawasaki Disease

Rosa Vona, Donatella Pietraforte, Lucrezia Gambardella, Alessandra Marchesi, Isabella Tarissi de Jacobis, Alberto Villani, Domenico Del Principe and Elisabetta Straface

Abstract

Kawasaki disease (KD) is a rare and often undiagnosed disease, at least in the western countries. Although its etiology remains unidentified, epidemiological features point to the role of infection and genetic predisposition. KD is characterized by an inflammatory acute febrile vasculitis. Coronary artery involvement is the most important complication of KD and may cause significant coronary stenosis resulting in ischemic heart disease. It has been demonstrated that the major risks in KD progression are the endothelial dysfunction and that systemic oxidative stress together with premature aging of red blood cells and alteration of platelet homeostasis, could play a critical role in the cardiovascular complications associated with KD. This chapter will focus on the role of oxidative stress in endothelial damage and on circulating blood cells of KD patients.

Keywords: etiology, oxidative stress, inflammation, biomarkers, red blood cells, platelets

1. Introduction

Kawasaki disease (KD) is an inflammatory acute febrile vasculitis that can also lead to coronary artery weakening, aneurysm formation, and myocardial infarction. The incidence of this disease varies considerably between ethnic groups: in Asians are up to 20 times higher than Caucasians. KD is most prominently recognized in Japan, Korea, and Taiwan, reflecting increased genetic susceptibility among Asian populations. The highest incidence is reported in Japan: about 90 per 100,000 [1, 2]. Although nearly 50 years have passed from the first description, the etiology of KD remains a mystery. Since the incidence of the disease is high among Japanese people, it can be speculated that this people may have some sort of genetic characteristic that leaves them susceptible to KD. In addition, both clinical and epidemiological findings strongly suggest that some infectious agent or bacterial super-antigenic toxin can play a pathogenetic role in genetically susceptible individuals [3]. Despite KD patients in the acute phase receive high-dose intravenous immunoglobulin (IVIG) and aspirin therapy, up to 5% of those affected will develop coronary aneurysms, predisposing them to thrombotic complications that could result in atherosclerosis, myocardial infarction, and/or death [4]. In fact, risk factors for the development of atherosclerosis such as C-reactive protein (CRP), oxidative stress (OS), and inflammatory cytokines, are increased in the acute phase of KD [5]. Moreover, in the acute phase of the disease, often patients undergo thrombocytosis that can exert a pathogenic role in the cardiovascular complications that characterize KD. However, in KD progression, the major risk is endothelial injury and coronary artery weakening, favoring the formation of aneurysms in 1:5 untreated children with KD as well as myocardial infarction, ischemic heart, and sudden death [6]. OS linked to inflammation that characterizes KD disease, has recently been included among the potentially useful diagnostic biomarkers in the vasculature of KD [7]. Several lines of evidence suggest that in KD patients, systemic OS may promote: (i) endothelial dysfunction through increased production of oxygen- and nitrogen-derived species (ROS/RNS); (ii) alter red blood cell (RBC) homeostasis, resulting in a sort of premature aging in these circulating cells that could lead to anemia and formation of blood clots; and (iii) stimulate platelet functions and defective platelet apoptosis program, resulting in thrombocytosis that can exert a pathogenetic role in the cardiovascular complications occurring in KD [8].

2. Kawasaki disease etiology

The etiology of KD remains one of the major mysteries in the field of Pediatrics, and no specific biological markers for diagnostic testing have been characterized to date. A large body of clinical, epidemiologic, immunologic, pathologic, and ultrastructural evidence suggests that environmental factors or infectious agents induce an intense inflammatory host response in genetically susceptible individuals [3]. The clinical findings of conjunctival injection, oral and pharyngeal erythema, cervical adenopathy, and rash, observed in patients with KD, are very similar to those observed in other pediatric infections acquired by the respiratory route.

2.1 Infections

Even if not confirmed, many published reports implicate a number of bacterial or viral pathogens such as *Staphylococcus*, *Streptococcus*, Adenovirus, human herpes virus 6 (HHV-6), Epstein Barr virus (HBV), human T-lymphotropic virus (HTLV), coronavirus and human bocavirus (HBoV) [9–19]. *Staphylococci and Streptococci* release exotoxins, known as super-antigens that promote the activation of a large numbers of T helper (Th) cells (5–20% of T cell clones) leading to an extensive immunological reaction [20]. Matsubara and collaborators state that toxic shock syndrome toxin-1 (TSST-1), Streptococcal Pyogenic Exotoxin A or C (SPEA or SPEC), and Staphylococcal Enterotoxin A or B (SEA or SEB) may act as super-antigens that could stimulate the immune system and result in KD [21]. However, despite an increase in: (i) anti-streptococcal SPEC antibodies in the sera of KD patients in acute phase [22] and (ii) anti-SPEC and -SPEA IgM found in the first few weeks following the illness [21], no significant differences in super-antigen antibody were found from some serological studies.

2.2 Immune dysregulation

Most investigators believe that derangement of the immune system and functional disorder of Th cells are the primary pathophysiologic features in patients with

KD [23]. Data analyses for KD show that abnormal immune responses to infectious agents play key roles in disease initiation. It has been reported that, in the acute phase of KD, viral or bacterial super-antigens act by binding to the Vβ region of the T cell receptor inducing a widespread immunological response and resulting in the release of pro-inflammatory cytokines such as tumor necrosis factor (TNF) α, interleukin (IL) 1β, 6, 8, interferon (IFN) γ, and chemokines, such as monocyte chemotactic protein-1 (MCP-1) [3, 24]. In fact, it has been found that serum levels of some cytokines, such as IL-6, IL-20, TNF-α, and IFN-γ increase significantly before IVIG treatment and that levels of IL-6, IL-10, and IFN-γ decreased rapidly after treatment [25]. Moreover, studies in a murine systemic vasculitis, induced by *Candida albicans* extract, emphasize a relationship between the development of vasculitis and the overexpression of pro-inflammatory cytokines, such as TNF-α and IL-6 [26, 27]. The activation of the immune system and the cascade of inflammatory factors are considered as important features of KD. In fact, Th cells, mononuclear cells, macrophages and plasma cells, with a smaller number of neutrophils, are observed in various organ tissues of fatal cases of acute KD [23]. When activated, T helper cells mainly differentiate into two functionally distinct subsets, Th1 and Th2 cells. Th1 cells play an important role in cellular immunity by secreting IL-2 and IFN-γ, while Th2 cells involve the development of antibody-producing B cells via the secretion of IL-4, IL-5, IL-6, and IL-10. Some of these cytokines play an important role in the progression from systemic activation of the immune system to local inflammation in coronary vessels. Recently, it has been demonstrated that KD patients may be non-responsive to IVIG when, after IVIG treatment, the serum levels of IL-6 and IL-10 decrease slowly and the levels of IL-4 and TNF-α increase [25]. Although activation of the immune system and production of various cytokines have both been reported in patients with KD, the role of T cells and the functional state of Th1 and Th2 cells in KD are still not fully understood. Moreover, an imbalance between the line Th 17 (Th17) and regulatory T (Treg) cells has been described in the peripheral blood from patients with KD [28]. Th17 cells have been identified as inflammation regulators via production of distinct cytokines, such as interleukin IL-17. Conversely, to the Treg cells expressing FOXP3 has been attributed an anti-inflammatory role via production of anti-inflammatory cytokines, for example, IL-10 and TGF-β1 [29]. T helper cells involved in KD etiology are listed in **Figure 1** and **Table 1**.

2.3 Genetics

For decades, researchers attempted to identify candidate genes conferring susceptibility to the KD. In particular, studies on genes related to innate and acquired immune functions or to vascular remodeling, have been conducted [30]. Genes for analyses were selected based on the information of their known function or role in the disease pathophysiology. Initial genetic studies were focused on human leukocyte antigen (HLA) genes, located at chromosome 6p21.3, that encode the protein on the cell-surface antigen-presenting proteins, involved in the regulation of the immune system. The roles of HLA genes have been investigated in several immune-mediated vascular diseases, including KD. The results of such studies vary depending on the ethnic group studied. A recent genome-wide association study demonstrated the significant association of HLA class II region (HLA-DQB2-DOB) with KD in a Japanese population [31]. A genome-wide association study conducted in a Korean population demonstrated a significant association with KD of the HLA class I locus that contains the HLA-B and HLA-C genes [32]. These studies suggest that either HLA class I or class II may be associated with KD and play a role in KD pathogenesis. Several reports show associations between

KD and specific HLA genotypes including HLA-B54 in a Japanese population [33], HLA-B51 in Caucasian populations [34], HLA-B35, -B75, and -Cw09 in Korean [35], and the major histocompatibility complex class I chain-related gene A (MICA) genes in southern Chinese [36]. Genome-wide association studies (GWASs) have identified several susceptibility genes associated with KD, including *CD40L*, *HLA-E*, *BLK*, and *FCGR2A* [37–40]. CD40L gene, located on Xq26, is known to induce endothelial cells to produce cell adhesion molecules and chemokines. Its expression has been found elevated on CD4+ Th cells and platelets during the acute-phase KD and in KD patients with coronary artery lesions (CALs)

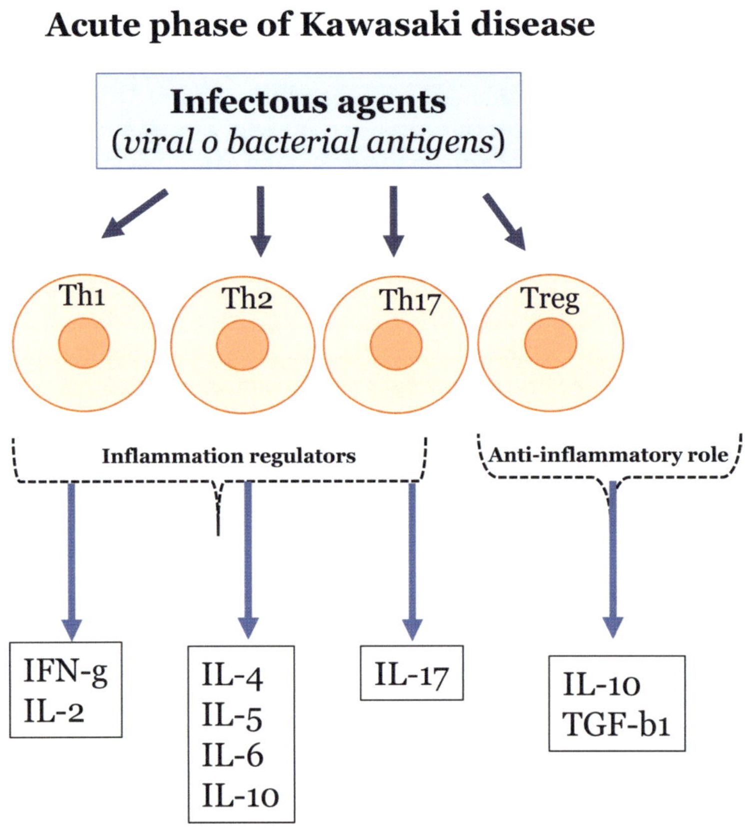

Figure 1.
T helper cells involved in KD etiology. In the acute phase of Kawasaki disease, viral or bacterial super-antigens induce immunological response resulting in the release of cytokines. Th cells are regulators of inflammation. T helper 1 cells secrete interferon g (IFN-g) and interleukin 2 (IL-2); T helper 2 cells secrete interleukin 4, 5, 6, and 10; T helper 17 cells secrete interleukin 17 (IL-17). Conversely, regulatory T cells (Treg) have an anti-inflammatory role via production of interleukin 10 (IL-10) and growth factor-beta (TGF-b).

T helper cells	Functions
Th1	Regulate cellular immunity by secreting IL-2 and IFN-γ
Th2	Regulate humoral immunity by secreting IL-4, IL-5, IL-6, and IL-10
Th17	Regulate inflammation by secreting IL-17
Treg	Anti-inflammatory role by the release of IL-10 and TGF-β1

Treg, regulatory T cells.

Table 1.
Immune dysregulation in KD: role of T helper cells.

[41]. HLA-E is a known ligand of CD94/natural killer cell receptor group 2-A (NKG2-A) and CD94/NKG2-C, which are expressed on natural killer cells [38]. Recent studies have shown that HLA-E has regulatory functions in both the innate and adaptive immune responses and that may have important implications in the pathogenesis of immune-mediated vascular diseases [42]. *BLK* is a Src family tyrosine kinase expressed primarily in the B-cell lineage and located on chromosome 8p22-23. During the acute and convalescent stages of KD, *BLK* expression correlates with the percentage of B cells in the peripheral blood mononuclear cells. Importantly, a decreased *BLK* expression in peripheral blood B cells may alter B cell function and predispose individuals to KD [43]. The *BLK* was significantly associated with KD susceptibility in Taiwanese and in Japanese populations [43]. *FCGR2A* gene is on chromosome 1q23 and encodes the FcγRIIA protein (CD32a), a member of a family of receptors for IgG (including the A, B, and C subunits of FcγRI and FcγRII and the A and B subunits of FcγRIII). This receptor is found on the surface of many immune cells, including natural killer cells, macrophages, and neutrophils, and it is involved in cellular activation and uptake of immune complexes [44]. The *FCGR2A* is associated with KD susceptibility in Korean and Asiatic populations [30]. Genes related to vasoactive or angiogenic molecules also can be considered as candidates for KD susceptibility or severity. Ohno and co-worker have shown that an up-regulation of vascular endothelial growth factor (VEGF) is involved in formation of coronary artery lesions (CALs) [45]. VEGF, expressed in various types of cells including leukocytes and vascular smooth muscle cells, binding to its receptor (VEGFR-1 and VEGFR-2) expressed on endothelial cells induces cell proliferation, survival, migration, and angiogenesis. Its ability to induce vascular hyper-permeability and chemotaxis of bone marrow-derived cells suggest significant roles of VEGF in inflammation [45]. Other candidate genes for KD are transforming growth factor-beta (TGF-β), because TGF-β-mediated T-cell activation and cardiovascular remodeling are important features of KD. This gene, located on chromosome 19q13.1, modulates the balance of pro-inflammatory/anti-inflammatory T cells through a complex set of interactions [46]. Genetic variations in the TGF-β pathway may lead to an imbalance of pro-inflammatory and regulatory T cells (Treg) by affecting the expression of the forkhead/winged helix transcription factor P3 (FOXP3) that is involved in the differentiation, function, and survival of CD4 + CD25+ regulatory T cells. Several studies demonstrated that in the peripheral circulation of KD patients, Treg cell numbers were reduced, and their function compromised [47]. Recently, in KD, 191 genes mainly implicated in inflammation and innate immune response and some signaling pathway such as platelet activation have been identified. Among these genes, *MAPK14* and *PHLPP1* were considered as the key functional genes that can distinguish KD from common infectious illness [48]. *MAPK14* is a gene that encodes p38α, a MAP kinases implicated in various cellular processes including proliferation, differentiation, transcription regulation, and development [49]. MPK14/P38 was found to significantly improve endothelial function and inflammation after vascular injury. *PHLPP1* encodes a protein that is a member of the Ser/Thr phosphatase family. Its upregulation in acute KD may reduce vascular injury by inactivating Akt and subsequent reducing the expression of NO [48]. Moreover, genetic polymorphisms of 1,4,5-trisphosphate 3-kinase C (ITPKC) and caspase 3 (CASP3) have been shown to associate with coronary artery lesions formation in both Japanese and Taiwanese populations of KD patients [50]. ITPKC is a gene located on chromosome 19q23 that acts as a negative regulator of T-cell activation. CASP3 is a gene located on chromosome 4q35, that is related to the apoptosis of immune cells [50]. Candidate genes in the KD etiology are listed in **Table 2**.

Candidate genes	Locus	Populations	Function
HLA-B54	6p21.3	Japanese	Regulation of the immune system
HLA-B51	6p21.3	Caucasian	Regulation of the immune system
HLA-B35	6p21.3	Corean	Regulation of the immune system
HLA-B75	6p21.3	Corean	Regulation of the immune system
HLA-E	6p21.3	Taiwanese	CAL formation
HLA-Cw09	6p21.3	Corean	Coronary complication
MICA	6p21.3	Southern Chinese	CAL formation
CD40L	Xq26	Taiwanese	CAL formation
BLK	8p22-23	Taiwanese and Japanese	Correlation with the % of B cells during KD
FCGR2A	1q23	Korean and Asiatic	Cellular activation and uptake of immune complexes
VEGF	6p12	Japanese	CAL formation
TGF-β	19q13.1	European	Modulates the balance of proinflammatory/anti-inflammatory T cells
MAPK14	6p21.31	Chinese	Autoimmunity-related vasculitis
PHLPP		Chinese	Reduce vascular injury
ITPKC	19q23	Taiwanese	Inactive T cells
CASP3	4q35	Taiwanese	Apoptosis in immune cells

HLA, human leukocyte antigen; MICA, major histocompatibility complex class I chain–related gene A; BLK, B-lymphoid tyrosine kinase; FCGR2A, Fc fragment of IgG receptor IIa; VEGF, vascular endothelial growth factor; TGF-β, transforming growth factor-beta; MAPK14, mitogen-activated protein kinase 14; PHLPP, PH domain leucine-rich repeat-containing protein phosphatase 1; ITPKC, 1,4,5-trisphosphate 3-kinase C; CASP3, caspase 3.

Table 2.
Candidate genes conferring susceptibility to the KD.

2.4 Environmental factors

Environmental factors, including socio-economic status and cultural habits in a society, affect the occurrence of infectious and autoimmune diseases. Recent studies suggest that environmental triggers, such as air pollution and extreme temperatures, may also serve as risk factors for KD [51]. Particulate matter and various gaseous pollutants, contained in the ambient air, have strong oxidizing property and the potential to induce KD through exaggerated inflammatory response, which is heavily involved in the pathophysiologic process of KD development [52]. Short-term exposure to air pollutants may damage endothelial cells, impair vascular function, stimulate systemic inflammation response, increase oxidative stress, and induce cardiac ischemia and repolarization abnormalities [52–54], consequently contributing to the development of KD. Moreover, from a time-stratified case-crossover study in Taiwan, evidence has been provided that exposure to ozone (O_3) may increase the risk of KD in children [55].

Recently, a study carried out on the Japanese population has found an association between higher household income, urbanization, and smaller family size at birth with increased KD incidence, which raises the hygiene hypothesis for the etiology of KD [56].

It has been reported that the human immune system and microbiota are trying to adapt to a changing environment. Gut microflora of infants were different according to ethnic groups, and the changing environment factors from industrialization may affect the distribution of gut microflora in infants [57]. Thus, it is very possible that normal flora also adjusts to a changing environment. Presently, the majority of data has found that the composition of the gut microbiota in KD patients differs from healthy subjects. Lee and co-workers have hypothesized that the immune system should lose tolerance to a part of the resident intestinal flora and that environmental factors, that is, a Western lifestyle or improved public hygiene systems, could transform the commensal flora into a pathogen one, as observed in different gastrointestinal disorders [58].

3. Implication of systemic oxidative stress in KD

It has been recognized that a systemic pro-oxidant state associated with inflammation can play a key role in the pathogenesis and progression of KD [59]. In support to this theory, experimental evidences showed increased concentration of oxidative stress-related biomarkers such as ROS/RNS, malondialdehyde (MDA), protein 3-nitrotyrosine, asymmetric dimethylarginine (ADMA), and myeloperoxidase (MPO). ROS/RNS are chemical heterogeneous molecules that include radical species, such as superoxide anion ($O_2^{\bullet-}$), hydroxyl radicals ($^{\bullet}OH$), and nitric oxide ($^{\bullet}NO$) and non-radical species such as hydrogen peroxide (H_2O_2,) and peroxynitrite (the product of the fast reaction between $O_2^{\bullet-}$ and $^{\bullet}NO$). Peroxynitrite-mediated oxidation includes its direct reaction with several cellular targets (CO_2, hemoproteins, and thiols), as well as indirect reaction, CO_2-dependent oxidations mediated by strong oxidizing radicals, such as $^{\bullet}NO_2$ and carbonate radical ($CO_3^{\bullet}$). The production of these oxidants is known to generate in blood a pro-oxidant status able to promote the occurrence of oxidative- and nitrative stress as well as redox imbalance leading to altered cell signaling and functions. These events may play a pathogenetic role in the cardiovascular complications often associated with KD [8]. As already mentioned, ROS/RNS generically can react with all the macromolecules of biological importance in cell and tissues, generating oxidative modification in lipids, DNA, and proteins that, in some cases, can be the footprint of the oxidant generated [60]. Malondialdehyde (MDA), the most investigated end-products of lipid peroxidation, is one of several low-molecular-weight end-products formed via the decomposition of certain primary and secondary lipid peroxidation products. It is a specific marker of omega-3 and omega-6 fatty acids peroxidation [61]. Increased serum levels of MDA were found in KD patients with coronary aneurysm associated with carotid intima-media thickening and stiffening [59]. Another marker of lipid peroxidation evaluated in KD patients is 8-isoprostaglandin F2α (8-iso-PG), a non-enzymatic oxidation product of arachidonic acid. Increased levels of 8-iso-PG have been measured in the urine from acute KD patients before IVIG therapy [62, 63]. Its increase reflects an enhanced endothelial dysfunction and correlates with cardiac dysfunction in acute KD [62]. Protein tyrosine nitration is an oxidative post-translational covalent modification of tyrosine residues consisting, in the addition of a nitro group ($-NO_2$) to the position 3, of the phenolic ring leading to the formation of 3-nitrotyrosine as an end-product [64]. It is a free-radical-mediated reaction induced by the one-electron oxidation of tyrosine residues to tyrosyl radical followed by its fast reaction with the nitrating agent $^{\bullet}NO_2$. In biological systems, 3-nitrotyrosine formation is mediated mainly by peroxynitrite-derived strong oxidants, such as $^{\bullet}OH$, $^{\bullet}NO_2$, $CO_3^{\bullet}$ [64]. In addition, 3-nitrotyrosine formation can be mediated by metals of heme-containing peroxidases in the presence of H_2O_2 and nitrite. The H_2O_2-genereted oxo-metal compounds (O = MnIV) and compounds I and II of heme-containing peroxidases, such as MPO, are highly heme oxidation state complexes able to oxidize tyrosine to tyrosyl radical, which in the presence of $^{\bullet}NO_2$, generate 3-nitrotyrosine [65, 66]. Protein tyrosine nitration is considered a hallmark of the reactions involving $^{\bullet}NO$-derived oxidant, that is, peroxynitrite and $^{\bullet}NO_2$, able to dramatically affect protein structure and function. Indeed, the occurrence of this oxidative modification leads to a loss- (superoxide dismutase, prostacyclin synthase, etc.) or to a gain-of-function (cytochrome c, protein kinase, glutathione S-transferase, etc.) of key macromolecules able to affect cell homeostasis and fate [64]. The well-established association of protein tyrosine nitration to several pathologies, such as cardiovascular disease, neurodegeneration, inflammation and cancer has made this protein modification not only a biomarker of RNS-derived oxidative stress *in vivo*, but also a predictor of disease onset and progression. MPO is a pro-oxidant enzyme released by activated poly-morphonuclear leukocytes that can promote the pro-inflammatory state by inducing the formation of RNS, 3-nitrotyrosine, and lipid peroxidation [65]. Furthermore, it can promote a blood pro-coagulant state favoring the binding of oxidized lipoproteins to the specific receptor present on the surface of platelets [67]. In this regard, 3-nitrotyrosine and MPO could play a pathogenetic role in the cardiovascular complications of KD and could be considered as biomarkers of inflammation in this disease. Indeed, elevated MPO levels were detected in acute KD patients before IVIG treatment [8]. It has been recognized that a persistent OS and an excessive ROS production play an integral role in the endothelial and smooth muscle dysfunction leading to the risk of premature arteriosclerosis in KD patients [68]. A longer duration of fever is associated with higher risk of oxidative stress-induced endothelial dysfunction [68]. ADMA, produced following the catabolism of proteins containing methylated arginine residues, is an endogenous inhibitor of the enzyme nitric oxide synthase (NOS), regulating the nitric oxide bioavailability. Many disease states, including cardiovascular diseases and diabetes, are associated with increased plasma levels of ADMA [69]. This compound could therefore play a crucial role in the pathogenesis of diseases associated with endothelial dysfunction, so that it has been proposed as a biomarker for cardiovascular risk. In plasma from KD patients, low levels of ADMA were detected before IVIG treatment and associated with coronary abnormalities [8, 70]. Moreover, it has been suggested that a pro-oxidant blood status could alter RBC homeostasis [71]. RBCs, under physiological conditions, represent the major components of blood antioxidant capacity and the cells with higher resistance to oxidative stress [71]. They exert a scavenging activity with a particular regard for ROS and for the species derived from nitric oxide, often overproduced in inflamed tissues. In fact, crossing inflamed areas can contribute to detoxify ROS and RNS "protecting" cells (e.g. endothelial cells). In contrast, when they cross a tissue where an intense production of ROS occurs, they may accumulate oxidative damage and become a source of reactive species capable of modifying the behavior and fate of endothelial cells [72]. In KD patients, alterations of RBCs, typically associated with oxidative imbalance, have been detected [8]. In particular, increased ROS levels and reduced intracellular total thiol content were measured in RBCs from KD patients before treatment with IVIG and aspirin. In addition, the appearance of RBCs with alterations typically associated with premature aging (e.g. glycophorin A and CD47 expression) or eryptosis (e.g. clustering of band 3 and increase of phosphatidylserine externalization) was observed. Glycophorin A (GA) is a glycoprotein widely expressed at the RBC surface that is downregulated during senescence. CD47 is an integrin-associated protein. Known as thrombospondin receptor, it acts as marker of self. Band 3 is an ion exchanger involved in RBC adhesion to endothelium. Phosphatidylserine (PS) is a phospholipid normally localized to the inner leaflet of the plasma membrane. During cell remodeling, it is externalized to the outer leaflet leading to RBC aging and death (eryptosis). Importantly, the appearance of aged and eryptotic RBCs in KD patients correlates with some clinical evaluations. In fact, it has been found that during the first 5 days of hospitalization, the number of RBCs, hemoglobin, mean corpuscular value, and hematocrit decrease significantly [8]. In addition, premature aging of RBCs, and their consequent removal from circulation, might be a risk factor for anemia: condition that can be found in KD patients. Furthermore, it has

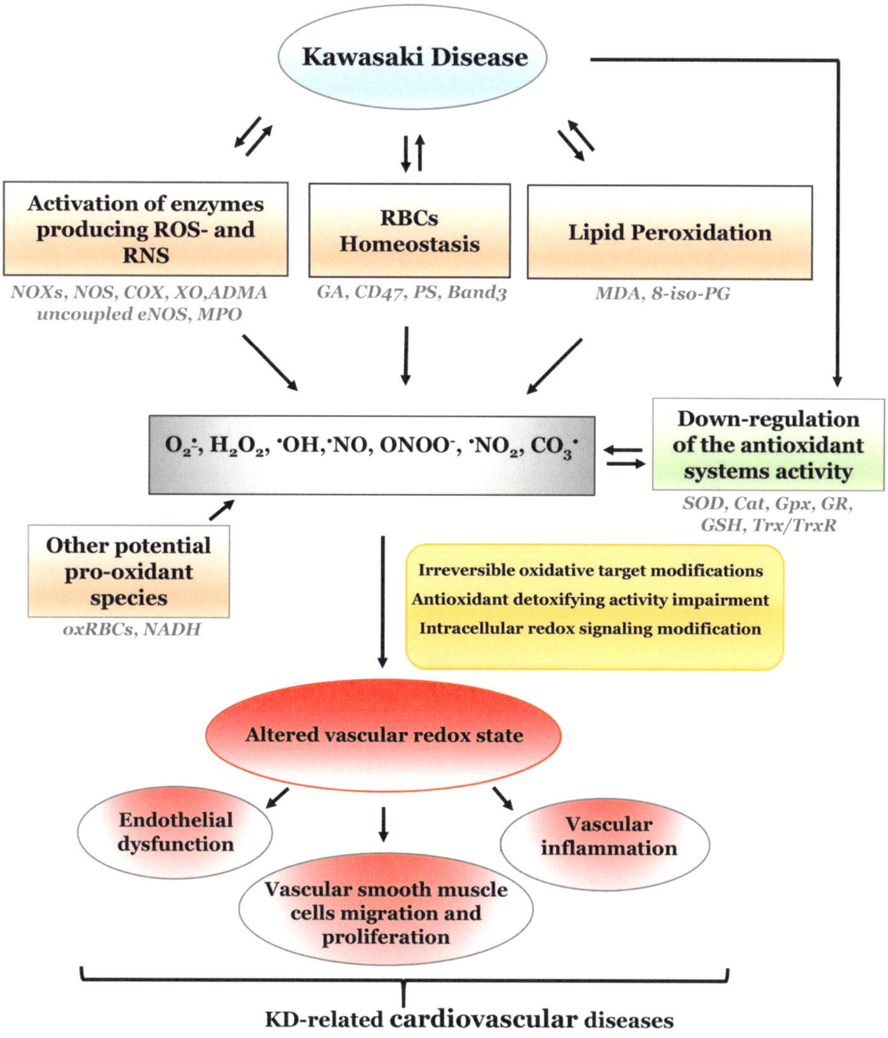

Figure 2.
Oxidative stress and vascular implications in Kawasaki disease. The Kawasaki disease (KD) is characterized by acute inflammation that has tissue oxidative stress as hallmark. This condition boosts the increase of reactive oxygen (ROS) and nitrogen (RNS) species formation in tissues and in the vasculature through the activation of the related producing enzymes in the cytosol. The cytosolic enzymes include the different isoforms of NADPH oxidase (NOXs), nitric oxide synthase (NOS), cyclooxygenase (COX), xanthine oxidase (OX), asymmetric dimethylarginine (ADMA), uncoupled endothelial NOS (eNOS), and myeloperoxidase (MPO). Other potential sources of ROS and RNS are the oxidized RBCs (oxRBCs) and lipid peroxidation. The Kawasaki disease is also characterized by the down-regulation of the antioxidant systems, including the depletion of GSH concentration and the decrease in the activity of the detoxifying enzymes, such as superoxide dismutase (SOD), catalase (Cat) glutathione peroxidase (Gpx), glutathione reductase (GR) and the couple constituted by thioredoxin (Trx) and thioredoxin reductase (TrxR). These conditions result in the irreversible accumulation of oxidation products in proteins, lipids, and sugars, which allow to the impairment of the intracellular redox signaling and detrimentally affect vascular biology by promoting vascular inflammation, endothelial dysfunction, and vascular remodeling. These alterations underlie the typical KD-associated cardiovascular complications, such as coronary artery weakening, aneurysm formation, and myocardial infarction.

been hypothesized that in KD patients, oxidative stress can alter platelet functions and platelet apoptosis program resulting in thrombocytosis that can exert a pathogenetic role in the cardiovascular complications [47]. This hypothesis is supported by the detection of markers of platelet activation, such as P-selectin shedding and PS externalization. P-selectin is a cell-adhesion molecule constitutively expressed in the α-granules of resting platelet. It translocates at the surface during platelet activation and subsequently released by a shedding phenomenon. Its release modulates leucocyte adhesion to both platelets and endothelial cells during inflammatory responses and thrombus formation [73]. PS externalization in platelets is usually associated with a sort of programmed cell death and correlated with their hyper-activation. In KD patients, before treatment with IVIG and aspirin, two different sub-populations of platelets have been identified: (i) annexin V positive platelets, characterized by a decreased mitochondrial membrane potential and

Plasmatic biomarkers	Specificity	Clinical findings in KD
ROS/RNS ($O_2^{\bullet-}$, $^{\bullet}OH$, $^{\bullet}NO$, H_2O_2)	Generates in blood a pro-oxidant status	Increased levels
MDA	Specific marker of omega-3 and omega-6 fatty acids peroxidation	Increased levels
Protein 3-nitrotyrosine	End-product of modification of tyrosine residues	Increased levels
ADMA	Endogenous inhibitor of the endothelial NOS. Regulates the NO bioavailability	Decreased levels
MPO	Pro-oxidant enzyme that can promote the pro-inflammatory state	Increased levels
RBC biomarkers		
Glycophorin A	Glycoprotein downregulated during RBC senescence	Down-regulated
CD47	Thrombospondin receptor that acts as a marker of self	Down-regulated
Band 3	Ion exchanger involved in RBC adhesion to endothelium	Down-regulated
PS externalization	Phospholipid, marker of RBC aging and death when externalized to the outer leaflet of the plasma membrane	Increased percentage of RBCs with externalized PS
Platelet biomarkers		
P-selectin	A cell-adhesion molecule that modulates leucocyte adhesion to both platelets and endothelial cells during inflammatory responses and thrombus formation	Shedding
PS externalization and loss of mitochondrial membrane potential	Biomarkers of pro-coagulant platelets	Detected
Mitochondrial membrane hyperpolarization without PS externalization	Biomarkers of potentially pro-coagulant platelets	Detected

ROS/RNS, oxygen- and nitrogen-derived species; $O_2^{\bullet}$, superoxide anion; $^{\bullet}OH$, hydroxyl radicals; $^{\bullet}NO$, nitric oxide; H_2O_2, hydrogen peroxide; MDA, malondialdehyde; ADMA, asymmetric dimethylarginine; MPO, myeloperoxidase; RBC, red blood cell; PS, phosphatidylserine; NOS, nitric oxide synthase.

Table 3.
Biomarkers of oxidative stress in KD.

defined as activated pro-coagulant platelets [74] and (ii) annexin V negative platelets, characterized by an increased mitochondrial membrane potential, prone to become pro-coagulant when in contact with adenosine diphosphate (ADP) and thromboxane, mediators normally released from activated platelets [75]. It has been hypothesized that in KD patients, activated pro-coagulant platelets could contribute to the increased thrombotic risk detected in these patients. Implication of oxidative stress in KD is depicted in **Figure 2**. Biomarkers of oxidative stress in KD patients are summarized in **Table 3**.

4. Conclusions

In this chapter, a complex framework of events contributing to the etiology of KD has been described. These include some type of bacterial or viral infection, genetic determinants, immune system as well as hematological alterations. Although epidemiological and clinical data suggest that KD may arise from an abnormal response to infectious diseases in genetically susceptible individuals, there are still many controversies about the etiology of KD. There is no agreement on KD-related infectious agents, and the immune mechanisms behind KD remaining only partially known. Only the basic research evaluating the pathogenic mechanisms of this disease will probably find new targets for identifying disease-modifying agents or therapies that are more specific. Moreover, in this chapter, we provided new lines of evidence supporting the hypothesis that systemic oxidative stress together with premature aging of RBCs and platelets could play a critical role in the cardiovascular risk observed in patients with KD.

Abbreviations

ADMA	asymmetric dimethylarginine
ADP	adenosine diphosphate
CALs	coronary artery lesions
CASP3	caspase 3
$CO_3^{\bullet}$	carbonate radical
CRP	C-reactive protein
GA	glycophorin A
GWASs	genome-wide association studies
H_2O_2	hydrogen peroxide
HBoV	coronavirus and human bocavirus
HBV	Epstein Barr virus
HHV-6	human herpes virus 6
HLA	human leukocyte antigen
HTLV	human T-lymphotropic virus
IFN-γ	Interferon γ
ITPKC	1,4,5-trisphosphate 3-kinase C
IVIG	intravenous immunoglobulin
KD	Kawasaki disease
MCP-1	monocyte chemotactic protein-1
MDA	malondialdehyde
MICA	major histocompatibility complex class I chain-related gene A
MPO	myeloperoxidase
NKG2-A	natural killer cell receptor group 2-A
$^{\bullet}NO$	nitric oxide

NOS	nitric oxide synthase
$O_2^{\bullet -}$	superoxide anion
$^{\bullet}OH$	hydroxyl radicals
OS	oxidative stress
PS	phosphatidylserine
RBC	red blood cell
RNS	reactive nitrogen species
ROS	reactive oxygen species
SEA	Staphylococcal Enterotoxin A
SEB	Staphylococcal Enterotoxin B
SPEA	Streptococcal Pyogenic Exotoxin A
SPEC	Streptococcal Pyogenic Exotoxin C
TGF-β	transforming growth factor-beta
TNF-α	tumor necrosis factor α
Treg	regulatory T cells
TSST-1	toxic shock syndrome toxin-1
VEGF	vascular endothelial growth factor

Author details

Rosa Vona[1], Donatella Pietraforte[2], Lucrezia Gambardella[1], Alessandra Marchesi[3], Isabella Tarissi de Jacobis[3], Alberto Villani[3], Domenico Del Principe[4] and Elisabetta Straface[1*]

1 Biomarkers Unit, Center for Gender-Specific Medicine, Istituto Superiore di Sanità, Rome, Italy

2 Core Facilities, Istituto Superiore di Sanità, Rome, Italy

3 General Pediatric and Infectious Disease Unit, Internal Care Department, Bambino Gesù Children's Hospital, Rome, Italy

4 Institute of Translational Pharmacology, CNR, Rome, Italy

*Address all correspondence to: elisabetta.straface@iss.it

References

[1] Makino N, Nakamura Y, Yashiro M, et al. Descriptive epidemiology of Kawasaki disease in Japan, 2011 2012: From the results of the 22nd nationwide survey. Journal of Epidemiology. 2015;**25**:239-245

[2] Bayers S, Shulman ST, Paller AS. Kawasaki disease: Part I. Diagnosis, clinical features, and pathogenesis. Journal of the American Academy of Dermatology. 2013;**69**:501-511

[3] Greco A, De Virgilio A, Rizzo MI, Tombolini M, Gallo A, Fusconi M, et al. Kawasaki disease: An evolving paradigm. Autoimmunity Reviews. 2015;**14**:703-709

[4] Paredes N, Mondal T, Brandão LR, Chan AK. Management of myocardial infarction in children with Kawasaki disease. Blood Coagulation & Fibrinolysis. 2010;**21**:620-631

[5] Fukazawa R, Ogawa S. Long-term prognosis of patients with Kawasaki disease: At risk for future atherosclerosis? Journal of Nippon Medical School. 2009;**76**:124-133

[6] Baker AL, Newburger JW. Kawasaki disease. Circulation. 2008;**118**:110-112

[7] Wenzel P, Kossmann S, Munzel T, Daiber A. Redox regulation of cardiovascular inflammation immunomodulatory function of mitochondrial and Nox-derived reactive oxygen and nitrogen species. Free Radical Biology & Medicine. 2017;**109**:48-60

[8] Straface E, Marchesi A, Gambardella L, Metere A, Tarissi de Jacobis I, Viora M, et al. Does oxidative stress play a critical role in cardiovascular complications of Kawasaki disease? Antioxidants & Redox Signaling. 2012;**17**:1441-1446

[9] Hall M, Hoyt L, Ferrieri P, Schlievert PM, Jenson HB. Kawasaki syndrome-like illness associated with infection caused by enterotoxin B-secreting *Staphylococcus aureus*. Clinical Infectious Diseases. 1999;**29**:586-589

[10] Shinomiya N, Takeda T, Kuratsuji T, Takagi K, Kosaka T, Tatsuzawa O, et al. Variant Streptococcus sanguis as an etiological agent of Kawasaki disease. Progress in Clinical and Biological Research. 1987;**250**:571-572

[11] Kikuta H, Nakanishi M, Ishikawa N, Konno M, Matsumoto S. Detection of Epstein-Barr virus sequences in patients with Kawasaki disease by means of the polymerase chain reaction. Intervirology. 1992;**33**:1-335

[12] Anderson DG, Warner G, Barlow E. Kawasaki disease associated with streptococcal infection within a family. Journal of Paediatrics and Child Health. 1995;**31**:355-357

[13] Embil JA, McFarlane ES, Murphy DM, Krause VW, Stewart HB. Adenovirus type 2 isolated from a patient with fatal Kawasaki disease. Canadian Medical Association Journal. 1985;**132**:1400

[14] Okano M, Luka J, Thiele GM, Sakiyama Y, Matsumoto S, Purtilo DT. Human herpesvirus 6 infection and Kawasaki disease. Journal of Clinical Microbiology. 1989;**27**:2379-2380

[15] Okano M. Kawasaki disease and human lymphotropic virus infection. Current Medical Research and Opinion. 1999;**15**:129-134

[16] Holman RC, Belay ED, Clarke MJ, Kaufman SF, Schonberger LB. Kawasaki syndrome among American Indian and Alaska Native children, 1980 through 1995. The Pediatric Infectious Disease Journal. 1999;**18**:451-455

[17] Principi N, Bosis S, Esposito S. Effects of coronavirus infections in children. Emerging Infectious Diseases. 2010;**16**:183-188

[18] Rowley AH, Baker SC, Shulman ST, et al. Ultrastructural, immunofluorescence, and RNA evidence support the hypothesis of a "new" virus associated with Kawasaki disease. The Journal of Infectious Diseases. 2011;**203**:1021-1030

[19] Catalano-Pons C, Giraud C, Rozenberg F, Meritet JF, Lebon P, Gendrel D. Detection of human bocavirus in children with Kawasaki disease. Clinical Microbiology and Infection. 2007;**13**:1220-1222

[20] Proft T, Fraser JD. Bacterial superantigens. Clinical and Experimental Immunology. 2003;**133**:299-306

[21] Matsubara K, Fukaya T, Miwa K, et al. Development of serum IgM antibodies against superantigens of *Staphylococcus aureus* and *Streptococcus pyogenes* in Kawasaki disease. Clinical and Experimental Immunology. 2006;**143**:427-434

[22] Yoshioka T, Matsutani T, Toyosaki-Maeda T, et al. Relation of streptococcal pyrogenic exotoxin C as a causative superantigen for Kawasaki disease. Pediatric Research. 2003;**53**:403-410

[23] Lv YW, Wang J, Sun L, Zhang JM, Cao L, Ding YY, et al. Understanding the pathogenesis of Kawasaki disease by network and pathway analysis. Computational and Mathematical Methods in Medicine. 2013;**2013**:989307

[24] Meissner HC, Leung DY. Superantigens, conventional antigens and the etiology of Kawasaki syndrome. The Pediatric Infectious Disease Journal. 2000;**19**:91-94

[25] Wang Y, Wang W, Gong F, Fu S, Zhang Q, Hu J, et al. Evaluation of intravenous immunoglobulin resistance and coronary artery lesions in relation to Th1/Th2 cytokine profiles in patients with Kawasaki disease. Arthritis and Rheumatism. 2013;**65**:805-814

[26] Takahashi K, Oharaseki T, Wakayama M, Yokouchi Y, Naoe S, Murata H. Histopathological features of murine systemic vasculitis caused by *Candida albicans* extract—An animal model of Kawasaki disease. Inflammation Research. 2004;**53**:72-77

[27] Miura NN, Komai M, Adachi Y, Osada N, Kameoka Y, Suzuki K, et al. IL-10 is a negative regulatory factor of CAWS-vasculitis in CBA/J mice as assessed by comparison with Bruton's tyrosine kinase-deficient CBA/N mice. Journal of Immunology. 2009;**183**:3417-3424

[28] Jia S, Li C, Wang G, et al. The T helper type 17/regulatory T cell imbalance in patients with acute Kawasaki disease. Clinical and Experimental Immunology. 2010;**162**:131-137

[29] Workman CJ, Szymczak-Workman AL, Collison LW, Pillai MR, Vignali DA. The development and function of regulatory T cells. Cellular and Molecular Life Sciences. 2009;**66**:2603-2622

[30] Onouchi Y. Molecular genetics of Kawasaki disease. Pediatric Research. 2009;**65**:46R-54R

[31] Onouchi Y, Ozaki K, Burns JC, Shimizu C, Terai M, Hamada H, et al. A genome-wide association study identifies three new risk loci for Kawasaki disease. Nature Genetics. 2012;**44**:517-521

[32] Kim JJ, Yun SW, Yu JJ, Yoon KL, Lee KY, Kil HR, et al. A genome-wide association analysis identifies NMNAT2 and HCP5 as susceptibility loci for Kawasaki disease. Journal of Human Genetics. 2017;**62**:1023-1029

[33] Kato S, Kimura M, Tsuji K, Kusakawa S, Asai T, Juji T, et al. HLA antigens in Kawasaki disease. Pediatrics. 1978;**61**:252

[34] Kaslow RA, Bailowitz A, Lin FY, Koslowe P, Simonis T, Israel E. Association of epidemic Kawasaki syndrome with the HLA-A2, B44, Cw5 antigen combination. Arthritis and Rheumatism. 1985;**28**:938

[35] Oh JH, Han JW, Lee SJ, Lee KY, Suh BK, Koh DK, et al. Polymorphisms of human leukocyte antigen genes in Korean children with Kawasaki disease. Pediatric Cardiology. 2008;**29**:402

[36] Huang Y, Lee YJ, Chen MR, Hsu CH, Lin SP, Sung TC, et al. Polymorphism of transmembrane region of MICA gene and Kawasaki disease. Experimental and Clinical Immunogenetics. 2000;**17**:130-137

[37] Onouchi Y, Onoue S, Tamari M, Wakui K, Fukushima Y, Yashiro M, et al. CD40 ligand gene and Kawasaki disease. European Journal of Human Genetics. 2004;**12**:1062-1068

[38] Lin YJ, Wan L, Wu JY, Sheu JJ, Lin CW, Lan YC, et al. HLA-E gene polymorphism associated with susceptibility to Kawasaki disease and formation of coronary artery aneurysms. Arthritis and Rheumatism. 2009;**60**:604-610

[39] Wang W, Lou J, Lu XZ, Qi YQ, Shen N, Zhong R, et al. 8p22-23-rs2254546 as a susceptibility locus for Kawasaki disease: A case-control study and a meta-analysis. Scientific Reports. 2014;**4**:4247

[40] Khor CC, Davila S, Breunis WB, Lee YC, Shimizu C, Wright VJ, et al. Genome-wide association study identifies FCGR2A as a susceptibility locus for Kawasaki disease. Nature Genetics. 2011;**43**(12):1241-1246

[41] Wang CL, Wu YT, Liu CA, et al. Expression of CD40 ligand on CD4+ T-cells and platelets correlated to the coronary artery lesion and disease progress in Kawasaki disease. Pediatrics. 2003;**111**:E140-E147

[42] Coupel S, Moreau A, Hamidou M, Horejsi V, Soulillou JP, Charreau B. Expression and release of soluble HLA-E is an immunoregulatory feature of endothelial cell activation. Blood. 2007;**109**:2806-2814

[43] Chang CJ, Kuo HC, Chang JS, et al. Replication and meta-analysis of GWAS identified susceptibility loci in Kawasaki disease confirm the importance of B lymphoid tyrosine kinase (BLK) in disease susceptibility. PLoS One. 2013;**8**:e72037

[44] Falcini F, Trapani S, Turchini S, Farsi A, Ermini M, Keser G, et al. Immunological findings in Kawasaki disease: An evaluation in a cohort of Italian children. Clinical and Experimental Rheumatology. 1997;**15**:685-689

[45] Ohno T, Igarashi H, Inoue K, Akazawa K, Joho K, Hara T. Serum vascular endothelial growth factor: A new predictive indicator for the occurrence of coronary artery lesions in Kawasaki disease. European Journal of Pediatrics. 2000;**159**:424-429

[46] Shimizu C, Jain S, Davila S, et al. Transforming growth factor-beta signaling pathway in patients with Kawasaki disease. Circulation. Cardiovascular Genetics. 2011;**4**:6-25

[47] Del Principe D, Pietraforte D, Gambardella L, Marchesi A, Tarissi de Jacobis I, Villani A, et al. Pathogenetic determinants in Kawasaki disease: The haematological point of view. Journal of Cellular and Molecular Medicine. 2017;**21**:632-639

[48] Chang D, Qian C, Li H, Feng H. Comprehensive analyses of DNA

methylation and gene expression profiles of Kawasaki disease. Journal of Cellular Biochemistry. 2019

[49] Bonney EA. Mapping out p38MAPK. American Journal of Reproductive Immunology. 2017;77:e12652

[50] Kuo HC, Hsu YW, Wu CM, Chen SH, Hung KS, Chang WP, et al. A replication study for association of ITPKC and CASP3 two-locus analysis in IVIG unresponsiveness and coronary artery lesion in Kawasaki disease. PLoS One. 2013;**8**:e69685

[51] Lin Z, Meng X, Chen R, Huang G, Ma X, Chen J, et al. Ambient air pollution, temperature and Kawasaki disease in Shanghai, China. Chemosphere. 2017;**186**:817-822

[52] Kelly FJ. Oxidative stress: Its role in air pollution and adverse health effects. Occupational and Environmental Medicine. 2003;**60**:612-616

[53] Brook RD, Rajagopalan S, Pope CA 3rd, Brook JR, Bhatnagar A, Diez-Roux AV, et al. Particulate matter air pollution and cardiovascular disease: An update to the scientific statement from the American Heart Association. Circulation. 2010;**121**:2331-2378

[54] Chen R, Zhao Z, Sun Q, Lin Z, Zhao A, Wang C, et al. Size-fractionated particulate air pollution and circulating biomarkers of inflammation, coagulation, and vasoconstriction in a panel of young adults. Epidemiology. 2015;**26**:328-336

[55] Jung CR, Chen WT, Lin YT, Hwang BF. Ambient air pollutant exposures and hospitalization for Kawasaki disease in Taiwan: A case-crossover study (2000-2010). Environmental Health Perspectives. 2017;**125**:670-676

[56] Fujiwara T, Shobugawa Y, Matsumoto K, Kawachi I. Association of early social environment with the onset of pediatric Kawasaki disease. Annals of Epidemiology. 2019;**29**:74-80

[57] Stearns JC, Zulyniak MA, de Souza RJ, Campbell NC, Fontes M, Shaikh M, et al. Ethnic and diet-related differences in the healthy infant microbiome. Genome Medicine. 2017;**9**:32

[58] Lee KY, Han JW, Lee JS. Kawasaki disease may be a hyper-immune reaction of genetically susceptible children to variants of normal environmental flora. Medical Hypotheses. 2007;**69**:642-651

[59] Cheung YF, Karmin O, Woo CW, Armstrong S, Siow YL, Chow PC, et al. Oxidative stress in children late after Kawasaki disease: Relationship with carotid atherosclerosis and stiffness. BMC Pediatrics. 2008;**8**:20

[60] Marrocco I, Altieri F, Peluso I. Measurement and clinical significance of biomarkers of oxidative stress in humans. Oxidative Medicine and Cellular Longevity. 2017;**2017**:6501046

[61] Signorini C, De Felice C, Durand T, Oger C, Galano JM, Leoncini S, et al. Isoprostanes and 4-hydroxy-2-nonenal: Markers or mediators of disease? Focus on Rett syndrome as a model of autism spectrum disorder. Oxidative Medicine and Cellular Longevity. 2013;**2013**:343824

[62] Takeuchi D, Saji T, Takatsuki S, Fujiwara M. Abnormal tissue Doppler images are associated with elevated plasma brain natriuretic peptide and increased oxidative stress in acute Kawasaki disease. Circulation Journal. 2007;**71**:357-362

[63] Takatsuki S, Ito Y, Takeuchi D, et al. IVIG reduced vascular oxidative stress in patients with Kawasaki disease. Circulation Journal. 2009;**73**:1315-1318

[64] Bartesaghi S, Radi R. Fundamentals on the biochemistry of peroxynitrite

and protein tyrosine nitration. Redox Biology. 2018;**14**:618-625

[65] Campolo N, Bartesaghi S, Radi R. Metal-catalyzed protein tyrosine nitration in biological systems. Redox Report. 2014;**19**:221-231

[66] van der Vliet A, Eiserich JP, Halliwell B, Cross CE. Formation of reactive nitrogen species during peroxidase-catalyzed oxidation of nitrite. A potential additional mechanism of nitric oxide-dependent toxicity. The Journal of Biological Chemistry. 1997;**272**:7617-7625

[67] Ueno K, Nomura Y, Morita Y, Eguchi T, Masuda K, Kawano Y. Circulating platelet-neutrophil aggregates play a significant role in Kawasaki disease. Circulation Journal. 2015;**79**:1349-1356

[68] Ishikawa T, Seki K. The association between oxidative stress and endothelial dysfunction in early childhood patients with Kawasaki disease. BMC Cardiovascular Disorders. 2018;**18**:30

[69] Rochette L, Lorin J, Zeller M, Guilland JC, Lorgis L, Cottin Y, et al. Nitric oxide synthase inhibition and oxidative stress in cardiovascular diseases: Possible therapeutic targets? Pharmacology & Therapeutics. 2013;**140**:239-257

[70] Huang YH, Tain YL, Lee CP, Kuo HC. Asymmetric and symmetric dimethylarginine are associated with coronary artery lesions in Kawasaki disease. The Journal of Pediatrics. 2014;**165**:295-299

[71] Minetti M, Malorni W. Redox control of red blood cell biology: The red blood cell as a target and source of prooxidant species. Antioxidants & Redox Signaling. 2006;**8**:1165-1169

[72] Buehler PW, Alayash AI. Oxygen sensing in the circulation: "Cross talk" between red blood cells and the vasculature. Antioxidants & Redox Signaling. 2004;**6**:1000-1010

[73] Furie B, Furie BC, Flaumenhaft R. A journey with platelet P-selectin: The molecular basis of granule secretion, signalling and cell adhesion. Thrombosis and Haemostasis. 2001;**86**:214-221

[74] Dale GL, Friese P. Bax activators potentiate coated-platelet formation. Journal of Thrombosis and Haemostasis. 2006;**4**:2664-2669

[75] Pietraforte D, Gambardella L, Marchesi A, Tarissi de Jacobis I, Villani A, Del Principe D, et al. Platelets in Kawasaki patients: Two different populations with different mitochondrial functions. International Journal of Cardiology. 2014;**172**:526-528

Chapter 4

Radiation-Generated ROS Induce Apoptosis via Mitochondrial

Sandra Claro, Alice Teixeira Ferreira and Maria Etsuko Miyamoto Oshiro

Abstract

Ionizing radiation (IR) causes an increase in intracellular calcium, alters contractility, and triggers apoptosis via the activation of PKCα and -ε in irradiated smooth muscle cells. The present study investigated the role of the mitochondria in these processes and characterized the proteins involved in IR-induced apoptosis. Intestinal smooth muscle cells were exposed to 10–50 Gy from a γ-source. ROS and H_2O_2 levels were measured with colourimetry and a DCFH-DA probe, and protein expression was analyzed by immunoblotting and immunofluorescence. The IR-induced generation of ROS was inhibited by glutathione, and apoptosis was mediated by the mitochondria via BAX, cytochrome c, and caspase 3. IR increased the expression of the cyclins A, B2, and E, and led to unbalanced cellular growth in an absorption dose-dependent manner. However, radiation did not induce alterations in the mitochondrial ultrastructure or in $K\Psi_{mito}$. In contrast, IR increased the nuclear expression of BAG-1, TNFα, PKCα, and -ε and cyclins A and E. In conclusion, IR triggers the activation of antiapoptotic proteins and enhances the risk of a second type of cancer in patients undergoing radiotherapy. In addition to increasing the radioresistance of cells, antiapoptotic proteins can also stimulate uncontrolled cell proliferation that culminates in mutagenesis.

Keywords: ROS, apoptosis, mitochondria, cyclins, smooth muscle

1. Introduction

The molecular pathways that induce and regulate apoptosis have been extensively studied [1, 2]. Apoptosis is characterized by the condensation of nuclear chromatin and blebbing of nuclear and cytoplasmic membranes, a process that leads to the formation of membrane-bound apoptotic bodies [3]. The proteolytic caspase cascade plays a central role in the apoptotic response, and proteins of the BCL-2 family are key checkpoints in the regulation of apoptosis [4, 5]. In healthy cells, the BCL-2 family is kept in an inactive form, with a complex distribution in the mitochondrial outer membrane (MOM), sarco/endoplasmic reticulum (SER), cytosol, and nuclear envelope [6].

The mitochondria also play a key role in Ca2+ homeostasis and oxidative stress [7]. Elevated intracellular calcium concentrations ($[Ca2+]_i$) do not seem to inhibit mitochondrial motility [8] but can lead to the opening of the mitochondrial transition pore (MTP) complex during the process of swelling, which is responsible in turn for the permeability of the MOM to large molecules and the collapse of the mitochondrial transmembrane electric potential ($K\Psi_{mito}$) [9]. Several studies have used tumor cells to investigate the molecular pathways involved in the regulation and triggering of apoptosis by ionizing radiation (IR) [10, 11], but IR is more effective in normal than neoplasic tissue; so it is important to minimize the exposure in it and to clarify the mechanisms involved in the cellular damage [12]. In addition, damage to healthy tissues due to IR used in cancer treatment is frequently associated with the appearance of a second cancer occurring in the radiated field or in its vicinity [13]. This event could be explained by remodeling of the molecular and cellular processes triggering a number of inter- and intracellular signaling cascades that regulate the progression of the cell cycle and cell survival [14–16].

The apoptotic pathway activated by IR is different from the extrinsic pathway activated by ligands and involves the generation of reactive oxygen species (ROS) and H_2O_2 [10, 17]. According to Orrenius [18], the enhanced ROS production regulates cellular metabolism, for the execution of the suicide program, by proteins released from the mitochondria. One of the factors involved in ROS-induced cell death is tumor necrosis factor alpha (TNFα) [15, 19], and mitochondria appear to participate in the production of this mediator. A number of hypotheses have been put forward to explain the mechanism by which TNFα cytotoxicity induces the intrinsic pathway [11]. Nevertheless, the mechanisms regulated by ROS is not totally clear, but our previous results described an increase in $[Ca2+]_i$ [20] and the activation of protein kinase C (PKCα and -ε) [21]. IR has not been directly demonstrated to affect proteins, including cyclins, cyclin-dependent kinases (CDKs), retinoblastoma protein (Rb), and E2F complex proteins [22–24], involved in the orchestration of the cell cycle. The goals of this study were to examine the extrinsic and intrinsic mechanisms involved in the apoptosis, and to investigate ROS and H_2O_2 generation and the mitochondria role under IR of intestinal smooth muscle cells from the guinea pig ileum.

1.1 Tissues and cell culture

Fragments of the longitudinal smooth muscle layer of the guinea pig ileum (LSMLGPI) were prepared as described previously [20, 21], and the IR exposure in tissue fragments and confluent cell cultures from the LSMLGPI were exposed to single dose of 10–50 Gy, emitted by a 60Co γ-source [25]. The samples were radiated with a total dose of 10–50 Gy, and were then maintained for 3 days in Dulbecco's Modified Eagle Medium (DMEM).

1.2 Colourimetry

The ROS level was measured in the homogenates using the fluorescent method described by Yagi [26].

The H_2O_2-induced lipid peroxidation (LP) was measured through the oxidation of Fe2+ in the presence of xylenol orange in a spectrophotometer [27].

1.3 Immunofluorescence analysis

a. The data were acquired and analyzed using a FACS Calibur flow cytometer and CellQuest software.

b. The cell death study was measured at 585/542 nm using the log or linear model in the FL-2 channel [20, 28].

c. To test if the generation of ROS contributes to apoptosis, some cultured cells were incubated with glutathione (GSH), 10^{-3} M reduced glutathione, and yeast glutathione reductase type II (0.08 units/mg protein) and then fixed and stained as described in section *b* [29].

d. The generation of H_2O_2 was measured with 2′7′-dichlorofluorescein diacetate (DCFH-DA, as described by Hasui [30]) in live cultured cells. The cells were suspended in PBS, mixed with 0.3 mM DCFH-DA at 37°C to allow the conversion to DCF, and analyzed at 570/530 nm in the FL-1 channel.

e. To measure the degree of unbalanced growth, cultured cells were detached and stained with acridine orange (AO) for the evaluation of the ratio of RNA content, according to Traganos [31].

f. The proteins involved in apoptosis were measured by immunofluorescence by specific antibodies, anti-: caspase 3, cyclin A, cyclin B2, cyclin E, PKCα, PKCε, TNFα, BAX, cytochrome c, BAG-1, BCL-2, and BCL-xL ([20, 32]).

g. The cyclins A, B2, and E, and the DNA content were analyzed by MODFIT 3.0 software as described by Gong [33].

1.4 Western blot analysis

The experimental procedure was performed as previously described [21] using LSMLGPI homogenates. The following antibodies were used, anti-: caspase 3, cyclin A and cyclin B2, cyclin E, BCL-xL, BAX, cytochrome c, and BCL-2.

1.5 Confocal microscopy

LSMLGPI cells were seeded onto glass coverslips and exposed to IR. The mitochondria were stained with a probe as described by Claro [20] in living cells.

For analysis of the $K\Psi_{mito}$, 0.5 μM $DiOC_6(3)$ was used in DMEM, in vivo. The fluorescence was measured between 546/500 nm. To confirm the mitochondrial accumulation of $DiOC_6(3)$, the cells were incubated with $K\Psi_{mito}$ inhibitors [34] for different periods of time.

1.6 Electron micrography

The cells were seeded as described by Claro [21], and were then radiated and fixed before being analyzed with a transmission electron microscope (1200 EXII, JEOL, Tokyo, Japan).

1.7 Fluorescence microscopy

Living cells were incubated with 2 μg/ml bisbenzimides diluted in DMEM and were analyzed between 461 and 350 nm, for DNA labeling.

1.8 Statistical analysis

Differences between irradiated and nonradiated groups were identified using the analysis of variance (ANOVA) of the unpaired Newman-Keuls tests (GraphPad Prism 5 software). Statistical significance was set at $P < 0.05$.

2. Results

We tested if LSMLGPI cells die by apoptosis in response to IR and observed that the maximum number of apoptotic bodies appeared 72 h following radiation with 10–50 Gy [21, 28]. The first step was to evaluate the effects of IR on the expression of cell-cycle proteins in LSMLGPI cells (**Figure 1**). In contrast to the cyclins B2 and E, the expression of cyclin A was unchanged at 24, 48, and 72 h postradiation. Subsequently, all proteins were analyzed at 24 h postradiation.

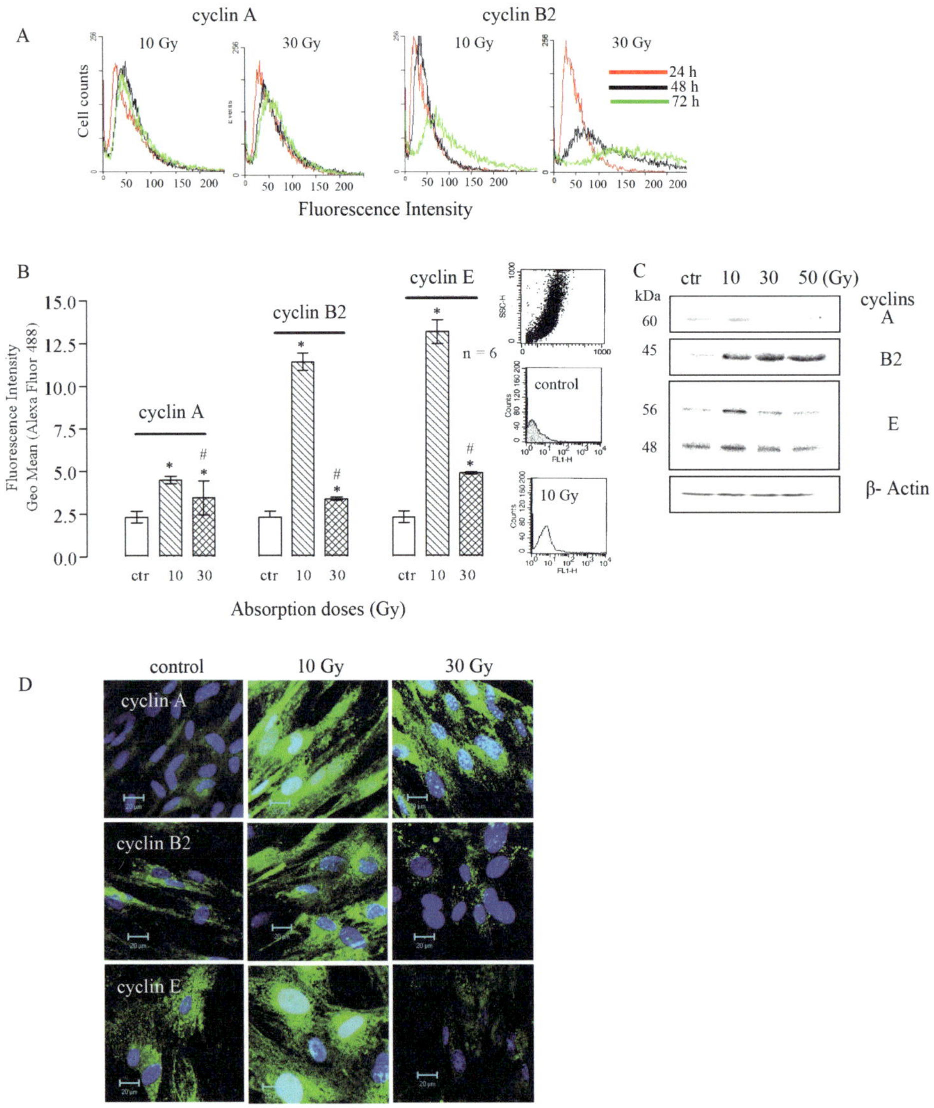

Figure 1.
*Effects of IR on the expression and localisation of cell cycle proteins. Cell cultures from LSMLGPI were fixed and labeled with specific primary and secondary antibodies. (A) Representative time-course histograms of activation of cyclins by IR. (B) Quantification of cells resuspended in PBS 24 h postradiation; besides, representative histograms of the acquisition data of relative cell size and analysis of fluorescence intensity distribution are shown. *P < 0.01 compared to control, #P < 0.01 compared to 10 Gy, Newman-Keuls test. Error bars indicate SEM. (C) Western blot analysis in whole-cell lysates demonstrating expression of cyclin a, B2 and E detected with appropriate antibodies. (D) Images of irradiated cells are representative of three independent experiments. Cyclin a and E co-localized with nucleus are light blue. Nuclear staining was done using DAPI (blue). Scale bar indicates 20 μm.*

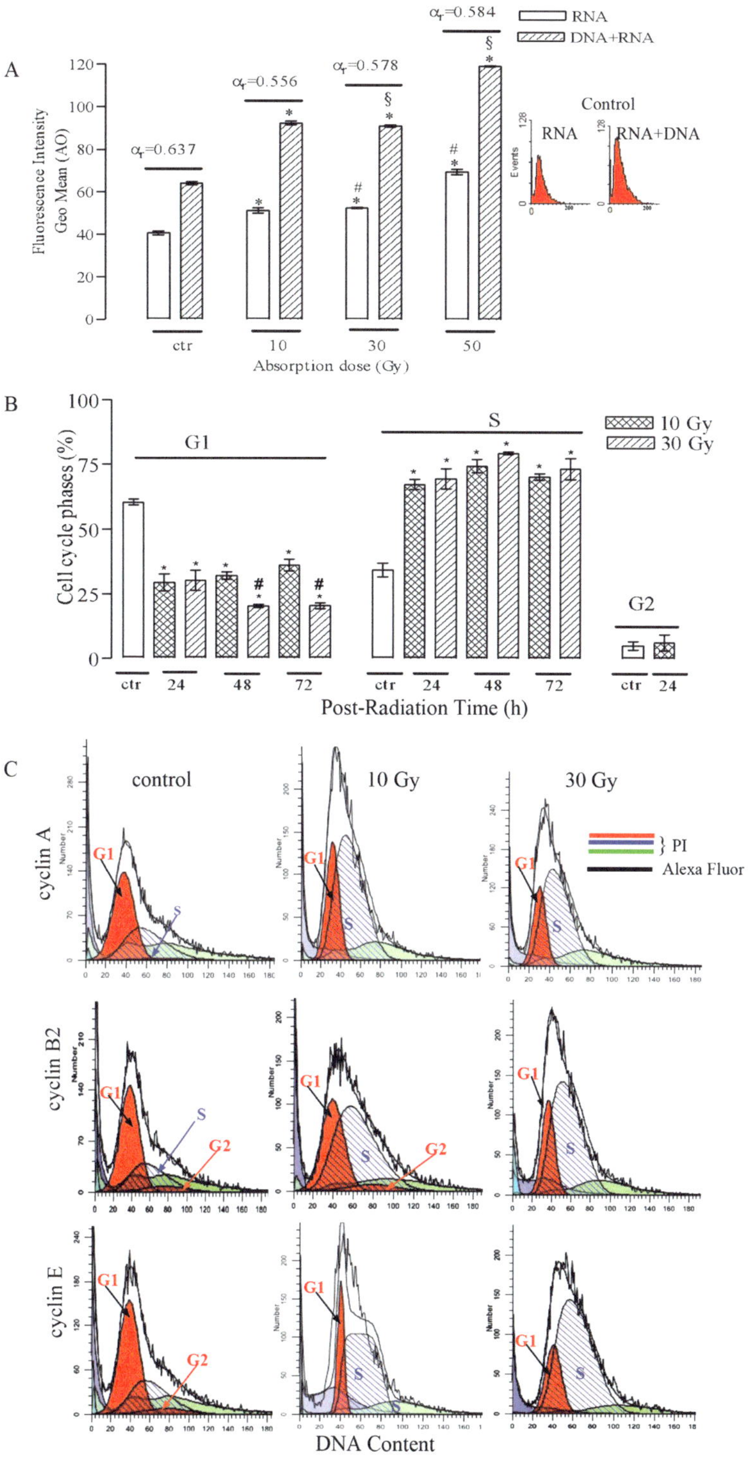

Figure 2.
*Effects of IR the on synthesis of RNA and DNA and on the cyclins in the cell cycle. (A) DNA, RNA and αr RNA/ (DNA + RNA) distribution 24 h postradiation; besides, representative histograms are shown. (B) Quantification of cell cycle phases by DNA content and analysis of G1, S, and G2 phases of cell cycle at different times of postradiation. (C) Scheme illustrating the analysis performed to estimate the cells expressing cyclins versus cell-cycle phases in measurements of cellular DNA content (PI) and the intensity of cyclins associated Alexa Fluor immunofluorescence analyzed by MODFIT 3.0 software. *P < 0.01 compared to control, #and §P < 0.01 compared to 10 and 30 Gy, respectively. Newman-Keuls test. Error bars indicate SEM.*

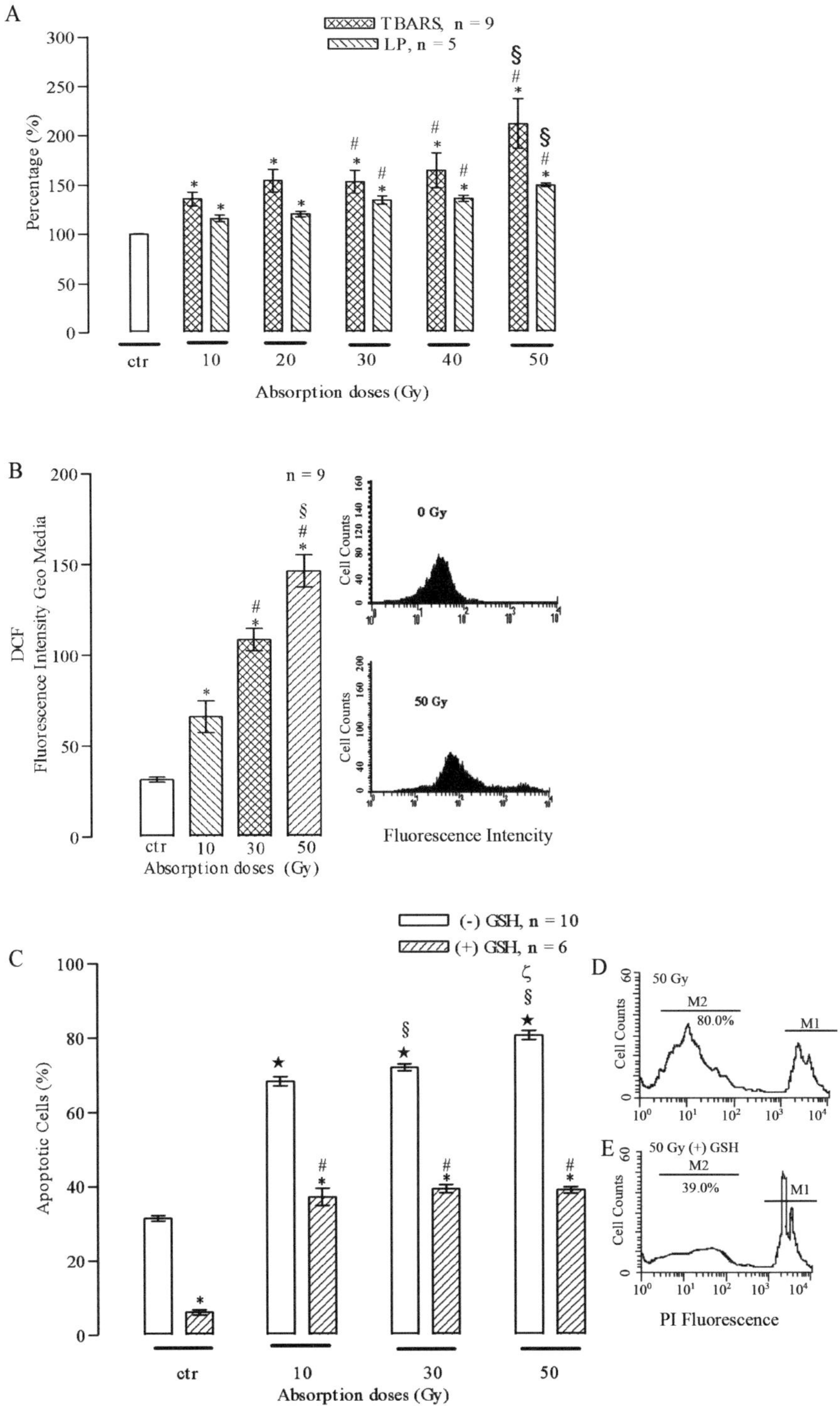

Figure 3.
*Measurements of IR-generated-ROS and H_2O_2. (A) TBARS and lipid peroxidation measured in homogenate of LSMLGPI via colourimetric assays. (B) Detection of intracellular H_2O_2 using DCFH-DA probe analyzed at flow cytometer, and (C) the representative histograms. *P < 0.01 compared to 0 Gy, #and §P < 0.01 compared to 10 and 30 Gy respectively. (D) Effect of glutathione on irradiated cells and fixed in 50% ethanol, and loaded with PI in the presence (+) or absence (−) of GSH, measured 72 h postradiation using flow cytometry and (E) representative histograms. *P < 0.01 indicates statistical difference between GSH-treated and untreated cells, #P < 0.01 compared to GSH-untreated control. *, § and ζP < 0.01 indicate statistical difference between untreated cells compared to control, 10 and 30 Gy, respectively. Newman-Keuls test. Error bars indicate SEM.*

Figure 2 correlates the changes in cyclin expression and the alteration of the cell cycle caused by IR. The α_r ratio of RNA to total nucleic acid content decreased in an absorption dose-dependent manner, and it visualizes nuclear content. The radiated population of cells did not divide because the G2 phase was arrested despite a significant increase in the accumulation of RNA and DNA during the S phase. Cyclins were continuously expressed during the cell cycle, however it was observed the G2 phase.

Figure 3 indicates that IR caused dose-dependent increases in the generation of thiobarbituric acid reactive substances (TBARS) and H_2O_2 with maximal ROS generation and a decrease in ROS levels. IR effects were suppressed by GSH, with a reduction in the number of cells in the M2 region. GSH reduced cell death independent on the dose of radiation, resulting in levels similar to those in control cells.

Apoptosis was assessed 24 h later by the binding of antibodies specific for BAX, cytochrome c, and caspase 3 (**Figure 4**). IR also increased the expression levels of BCL-xL and BCL-2, suggesting that these oncoproteins attempted to promote cell proliferation.

Figure 5 shows the stained apoptotic bodies and the localization of Bax, caspase 3, cytochrome c, Bcl-2, and BCL-xL.

Figure 6 proves that mitochondria presented no evidence of damage other than the appearance of several lysosomes. To prove that the mitochondria were healthy, various agents known to reduce the $K\Psi_{mito}$ were incubated with $DiOC_6(3)$, in living cells.

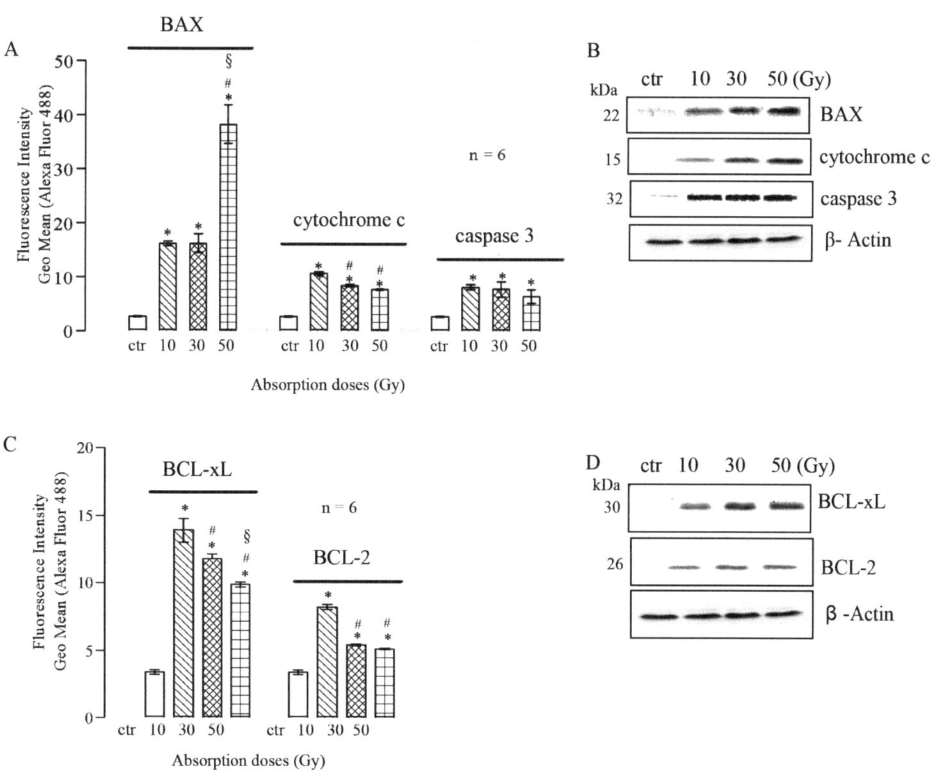

Figure 4.
*Effects of IR on (A) pro- and (C) antiapoptotic proteins of LSMLGPI cells measured in the flow cytometer 24 h postradiation. Cells were fixed, permeabilised and incubated with specific primary and secondary antibodies and resuspended in PBS. *P < 0.001 compared to control, #and §P < 0.01 compared to 10 and 30 Gy, respectively, Newman-Keuls test. Error bars indicate SEM. (B) and (D) western blot analyses demonstrating BAX, cytochrome c, caspase 3, BCL-xL, and BCL-2 expression, respectively.*

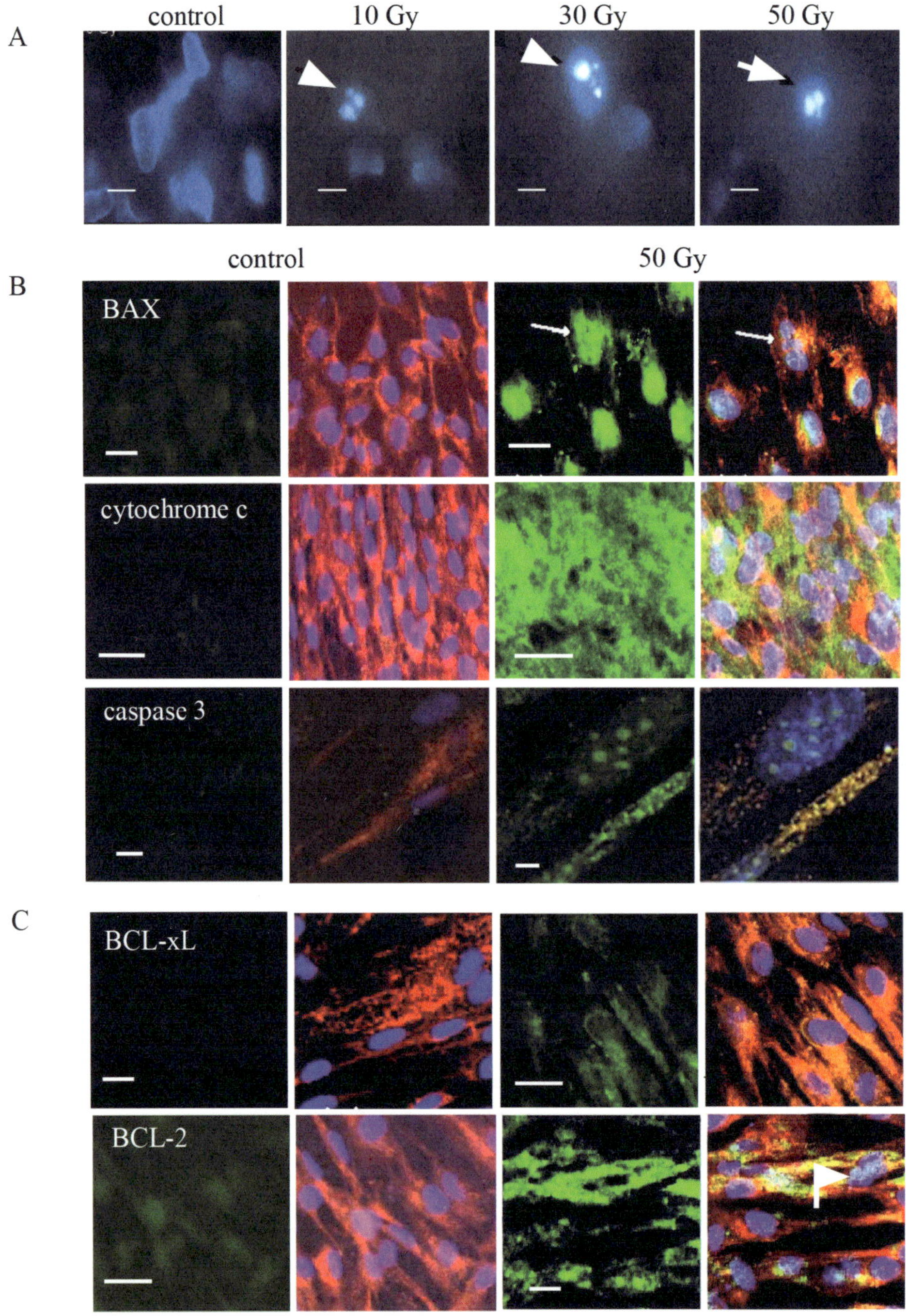

Figure 5.
Effects of IR on apoptotic proteins localisation 24 h postradiation. (A) Cell death by apoptosis is shown by apoptotic bodies formation in irradiated living cells labeled with 2 μg/ml Hoechst 33342 resuspended in cultured medium DMEM maintained at 37°C. Control cells exhibit low blue fluorescence, while irradiated cells exhibit high blue fluorescence and some apoptotic bodies (arrows). Images of irradiated cells present (B) proapoptotic and (C) antiapoptotic proteins with mitochondria stained with Mitotracker (red), and cells incubated with specific primary and secondary antibodies. Nuclear staining was done using DAPI (blue). Proteins co-localized with mitochondria are yellow. Arrows indicate apoptotic bodies. Images are representative of three independent experiments observed in confocal microscope. Scale bar indicates 20 μm.

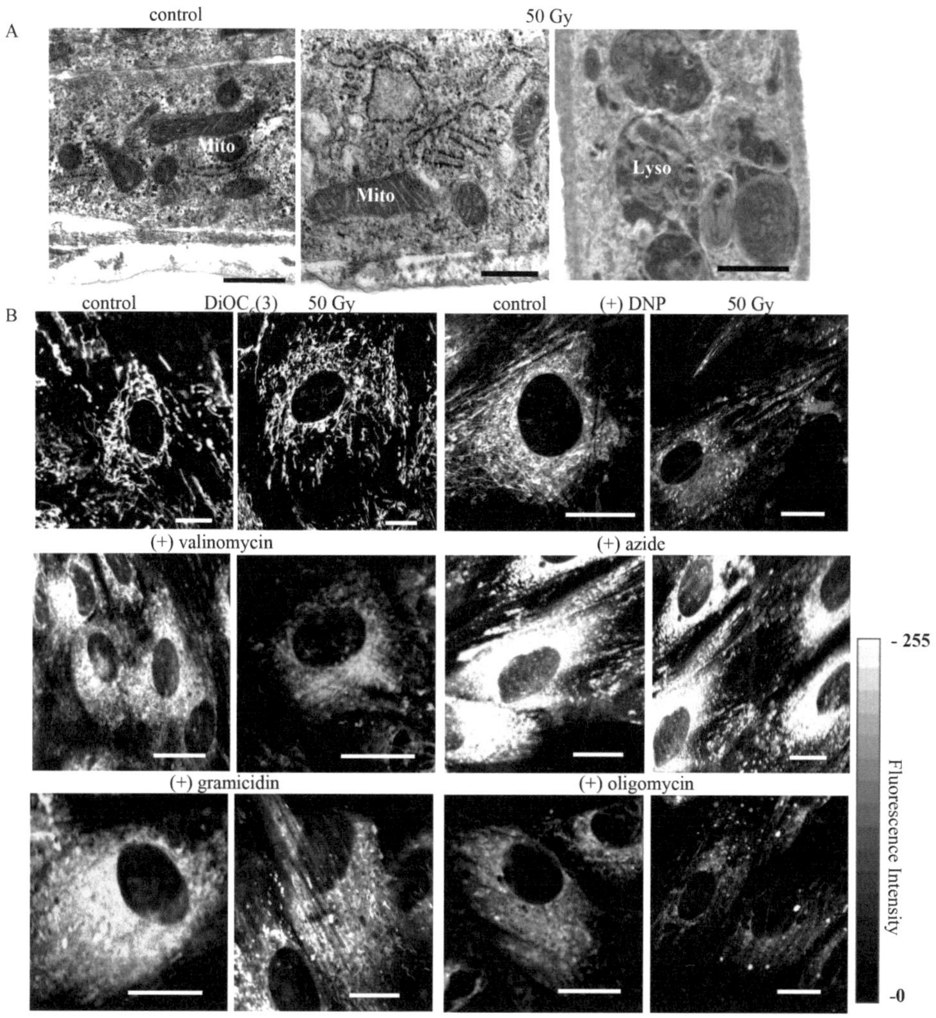

Figure 6.
Effect of IR on mitochondria in LSMLGPI cells cultures 72 h postradiation. (A) Electron microscopic analysis showing the mitochondria (Mito) with normal morphology scattered in the cytosol of control and lysosomes (Lyso); scale bar indicates 0.5 μm. (B) Confocal microscopy images in living cells loaded with $DiOC_6(3)$ and kept at 37°C. Cells were photographed before administration of ionophores, and after exposure to 4.5 nM valinomycin, 1 μM gramicidin, 1 mM DNP, 10 mM sodium azide, and 6.5 μM oligomycin; scale bar indicates 50 Zm. The figures are representative of three independent experiments.

The increased levels and activation/translocation of PKCα and -ε to the nucleus induced IR. Similarly, a large part of the TNFα was internalized and BAG-1 immunofluorescence appears next to the nucleus (**Figure** 7).

3. Discussion

IR generates ROS and H_2O_2 and promotes changes related to the expression and localization of cyclins, and in the cellular cycle phase distributions in a dose-dependent manner in LSMLGPI. Cyclins were continuously expressed during the cell cycle after treatment with IR; however, an arrest of the G2 phase and enhanced DNA replication at the initiation of the S phase occurred. The G2 phase is known

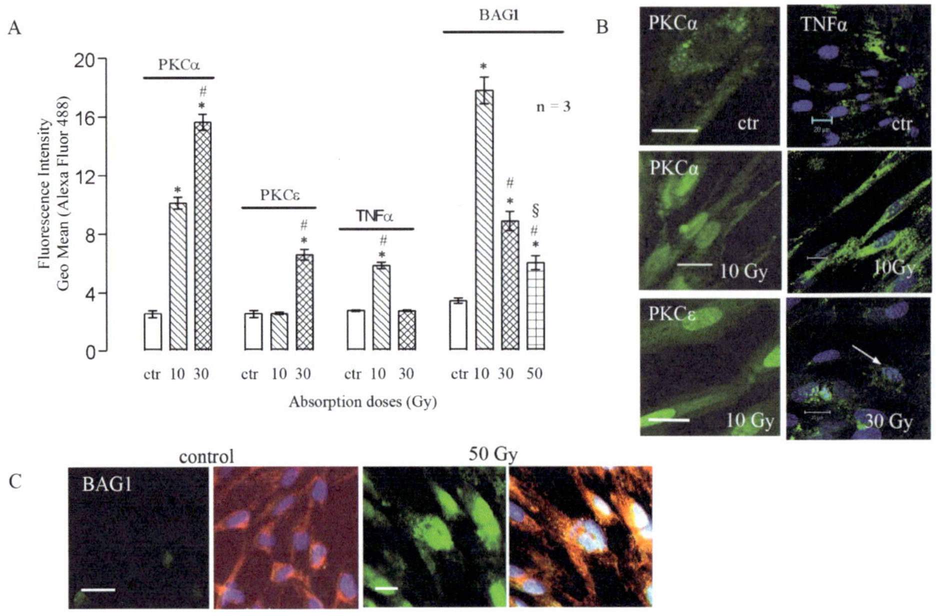

Figure 7.
Effects of IR on the expression and localisation of TNFα and BAG1, PKCα, and -ε, of LSMLGPI cell cultures 24 h postradiation. Cells were fixed and incubated with specific primary and secondary antibodies. (A) Quantification using flow cytometry in cells resuspended in PBS. $^{*}P < 0.01$ *compared to control,* $^{\#}$*and* $^{\S}P < 0.01$ *compared to 10 and 30 Gy, respectively, Newman-Keuls test. Error bars indicate SEM. Figures are representative of three independent experiments and present enhanced green fluorescence of (B) PKCα, PKCε, TNFα, and (C) BAG-1 co-localized with mitochondria that are yellow and with nucleus that are light blue (arrow shows apoptotic bodies). Nuclear staining was done using DAPI (blue). Scale bar indicates 20 μm.*

to be the most radiosensitive phase of the cell cycle, followed by the G1 phase [35]; thus, cells in the G2 phase did not continue to synthesize RNA or DNA. IR induced an excess of DNA in relation to RNA content. These results demonstrate that IR interferes in the cell-cycle distribution, but it does not cause cyclins degradation.

Cell death was effectively triggered by the activation and translocation of BAX to the mitochondria, resulting in cytochrome c release into the cytosol in an absorption dose-dependent manner. Ultrastructural changes and DNA fragmentation characteristics of apoptosis were also identified in vitro [21] and it was confirmed by Hoechst which stained the apoptotic bodies in living cells.

The BAX fluorescence intensity was increased next to the perinuclear region, with some co-localization with the MOM (yellow). Caspase 3 was overexpressed in the nucleus and co-localized with the mitochondria (yellow), and possible retention in the intermembrane space. We also observed caspase 3 localization in the nucleolus which is an atypical form. As cytochrome c mediates the activation of caspases via BAX disruption, we hypothesized that it might also induce the activation of antiapoptotic proteins. According to Edlich [36], activation of BCL-xL and BCL-2 increased the cellular resistance to death and could also cause the retrotranslocation of BAX to the cytosol, confirming our results. Our results demonstrated that there is more than one type of cellular response to IR, namely death or survival. The mitochondrial ultrastructure and function appeared normal in IR-induced apoptosis.

We have shown that IR causes apoptosis which is preceded by the activation of PKCα and -ε and suggests a role for the PKC-mediated pathway [21] and caspase 12 translocation to the cytosol [20]. We and other authors have shown that single absorption doses induce early reactions in normal smooth muscle cells, including protein breakage and the degradation of membrane phospholipids. However, ROS and H_2O_2 also cause DNA fragmentation and prevent the repair mechanisms elicited by sublethal damage [20, 21, 37]. ROS and H_2O_2 have been implicated in several mechanisms of cellular injury, including peroxidation of membrane phospholipids, which increases membrane permeability and leads to apoptosis ([38], pp. 196–208). In the present study, however, we observed that up to 50 Gy of IR led to cell death by apoptosis, despite the preservation of the plasma membrane. It is possible that H_2O_2, rather than ROS, can cross cell membranes rapidly and cause LP in small, discrete sites on smooth muscle membranes ([38], pp. 79–80). In contrast, ROS can mediate necrosis in neurons by the MTP pathway [18]. H_2O_2 is a weak oxidizing agent but can form hydroxyl radicals. These findings suggest that IR-generated ROS or H_2O_2 favors the internalization of TNFα. Several mechanisms may have protected the cells against injury in the presence of GSH, including the prevention of protein oxidation, the accumulation of H_2O_2 through its transformation in water ([38], pp. 10–21), the provision of a substrate for glutathione peroxidase, and the scavenging of hydroxyl radicals. Nevertheless, the most remarkable effect of GSH appears to be protection against alterations in the cell cycle ([38], pp. 247–251).

In fact, here, we show that high concentrations of ROS or H_2O_2 generated by IR were followed by the release of cytochrome c from the mitochondria into the cytosol. Several models of cytochrome c release have been proposed [2, 5], such as release through the MTP mega channel [39].

The mechanisms involving BAX, which is inserted into the MOM, may include the formation of channels, by oligomerization, and the preservation of mitochondrial membrane integrity [40]. Although we cannot discount the possible involvement of heterodimers among activated BCL-2, BCL-xL and BAG-1 proteins, there is no clear evidence that any of these have pore-forming activity [41].

The mitochondrial membranes were maintained intact in radiated cells, with similar fluorescence as the control cells, in which the electronegativity of the probe allowed its retention in the mitochondrial interior [34], $K\Psi_{mito}$ was maintained.

Our data indicate an intrinsic mechanism of IR-induced apoptosis. Moreover, this mechanism may be different in different types of mitochondria [15, 37].

Another potential repair mechanism is the decrease in the cellular ROS or H_2O_2 levels induced by BCL-2 [42]. This mechanism may also be activated by increased levels of antiapoptotic proteins BCL-xL and BAG-1. However, it has been suggested that BCL-2 survival factors are characteristic of cancer cell metabolism [43].

In addition to this survival pathways, that prevented cell death, we observed that BCL-2, BCL-xL, and BAG-1 were activated by direct IR and/or indirect via ROS or H_2O_2 action [44, 45]. Besides, the mitochondrial pattern can vary on different cells and it causes apoptosis that could be independent on the mitochondrial pathway [15, 37]. The radioresistance of mitochondria may be due to the action of natural antioxidants ([37, 38], pp. 97–98) and/or other compounds [46].

Increases in $[Ca2+]_i$ can potentiate the effects of ROS by enhancing LP [8, 14, 47]. ROS and increased $[Ca2+]_i$ have been shown to induce opening of the MTP, which triggers the mitochondrial of cell death [47]. It is noteworthy that mitochondria are located close to the SER, which sequestrates part of the Ca2+ released by these organelles, and this may affect the release of apoptotic and antiapoptotic factors from the SER [48–52]. The mitochondrial morphology may be altered by Ca2+ overload, with an increase in the MOM permeability culminating in the release of proapoptotic factors [8, 11]. However, our data demonstrated that the mitochondrial motility was maintained even in elevated $[Ca2+]_i$ after IR [20]. Increases of $[Ca2+]_i$ can also inhibit DNA and protein synthesis as well as nuclear transport, resulting in an accumulation of cells in the quiescent state (G0) [23]. In addition, $[Ca2+]_i$ up to 500 nM has been implicated in the regulation of the mammalian cell cycle during the early G1 phase and in the transition from the G1 to S phase [53]. Ca2+/calmodulin may also modulate the activity of cyclin-dependent kinases (CDK) and/or cyclin E [54]. In previous studies [20], we observed an increase in basal $[Ca^2+]_i$ cells was observed and it was suggested that IR causes modifications in the plasma membrane and/or in the sarco/endoplasmic reticulum, but the capacitative Ca2+ entry into radiated cells was reduced [55].

The cyclins A and E are constitutively nuclear proteins when involved in mitosis [14, 16]; nevertheless, in radiated cells, they leaked from the nucleus to the cytosol. The cyclin B2 complex appears to be localized predominantly in the SER [14, 16, 22, 23]. At the start of mitosis, cyclin B2 is rapidly transported into the nucleus [14]. An important fact to consider is that IR induced unbalanced growth [31]. Similar mechanism to Polavarapu [56] could be explained is the penetration of TNFα in the intestinal smooth muscle. According to our results, TNFα may penetrate the intracellular compartment through damage caused by lipid peroxidation in small, discrete sites of plasma membrane, since there is an ability of TNFα to form pores in biomembranes, or through the conventional receptor/lysosome route [46]. Also, activated TNFα can contribute to the apoptosis, as caused by ROS or H_2O_2. The increased TNFα expression in the cytosol could be explained by the presence of lysosomes in radiated cells, and we can infer that the TNFα was not subject to lysosomal autodigestion, since the mitochondrial membranes were preserved. TNFα can induce cell survival by the polymerization and depolymerization of actin filaments, which prevent the nuclear translocation of proapoptotic molecules and subsequently inhibit caspase 3 [57]. The activation involving ROS or H_2O_2 has been associated with the triggering of cell death modulated by TNFα [10, 15], through the activation of BAX or the protease cascade [58]. TNFα can also be involved in cell survival similar to IR models with higher doses [41]. In addition, we can infer that caspase 3 may enter into the MOM through membrane openings caused by activated BAX or TNFα [39, 59].

IR induces the formation of apoptotic bodies which will remain in the medium of cultured cells or they will be phagocytosed and digested by adjacent cells in the tissue [60]. Although DNA lesions induced by IR are lethal if not properly repaired, it is clear that membrane events may also contribute to radiation-induced apoptosis [61].

Our experiments demonstrated that radiation induced atypical activation of PKCα and -ε, and there is evidence that this may be related to a conservative regulation of cell cycle events, which act as a molecular link connecting signal transduction pathways and constituents of the cell-cycle machinery [62]. PKC participate in the control of G1 and G2/M, and PKCα and -ε may be regulators of the G1 phase and cause a delay in the G1/S transition, thereby halting DNA synthesis and contributing to cellular differentiation or death. In addition, we suggest that PKCα and -ε trigger cyclin activation and translocation to the nucleus, which occur through the C-terminal region [63]. The mechanism involved in the nuclear localization of PKCα and -ε after IR could be similar to that of PKCγ [63] but still remains to be determined. In contrast, the activation of PKCα and -ε may also have been induced by TNFα, with apoptosis triggered via activation of the TNF-receptor, in addition to elevated calcium, ROS and H_2O_2 levels [10, 15, 54]. PKCα and -ε may interact with the cyclins A, B2, and E in the mechanism of cellular survival, similar as the CDKs and PKC which have domains that may activate serine/threonine protein kinases [64, 65], in an atypical fashion. The involvement of PKCα and -ε activation in apoptosis has already been suggested [21].

We can speculate that cyclin E modulates PKCα and -ε when involved in the apoptosis. This possible involvement of PKCε would constitute a new finding, as currently it has only been associated with oncogenesis [66, 67]. Similar to TNFα, PKCε also contains an actin binding site, and its direct interaction with actin is essential for the invasion and metastasis of tumors grown in vitro or in vivo in the regulatory domain [66–68].

An important outcome of the complex network of events triggered by IR is the activation of antiapoptotic proteins in patients with cancer, and radiation therapy may lead to an increased risk of a second cancer [13]. In addition to their maleficent role in increasing radioresistance in normal cells, antiapoptotic proteins can stimulate uncontrolled cellular proliferation that culminates in carcinogenesis and mutagenesis [43]. Takayama et al. [69] identified BAG-1 and BCL-2 heterodimers that suppress apoptosis. Furthermore, BAG-1 overexpression is an important prognostic indicator of malignant tumors and may help to identify the metastatic potential of tumoral cells in vivo [70]. BCL-2 can alter the distribution of intracellular BAG-1, thereby changing the cancer risk [70]. Therefore, the overexpression of BCL-2, BCL-xL, and BAG-1 in normal cells may be a predictive indicator of carcinogenesis [69, 70]. In addition, PKCε is an important signaling molecule that influences the levels/activation of antiapoptotic proteins of the BCL-2 family and may regulate mitochondrial integrity, which is associated with cancer [71, 72]. However, the mechanism by which proteins of the BCL-2 family regulate cell death remains controversial. Our data suggest that not only apoptosis but also cellular repair mechanisms are activated in smooth muscle cells subjected to a low absorption dose.

Additionally, the expression level and localization of these proteins may be an important survival indicator in irradiated normal cells and may inform the prognosis of cancer patients undergoing radiotherapy.

Acknowledgements

The authors would like to thank Fundação de Amparo e Pesquisa do Estado de São Paulo (FAPESP); Federal University of São Paulo (UNIFESP); Edgar Paredes-Gamero, Soraya Smaili, Gustavo José Pareira and Renato de Arruda Mortara.

Author details

Sandra Claro*, Alice Teixeira Ferreira and Maria Etsuko Miyamoto Oshiro
Departamento de Biofísica, Federal University of São Paulo, São Paulo, Brazil

*Address all correspondence to: claro.sandra@unifesp.br

References

[1] Andersen JL, Kornbluth S. The tangled circuitry of metabolism and apoptosis. Molecular Cell. 2013;**49**:399-410

[2] Youle RJ, Strasser A. The Bcl-2 protein family: Opposing activities that mediate cell death. Nature Reviews. Molecular Cell Biology. 2008;**9**:47-59

[3] Kerr JR, Wyllie AH, Currie AR. Apoptosis: A basic biological phenomenon with wideranging implications in tissue kinetics. British Journal of Cancer. 1972;**26**:239-257

[4] Burlacu A. Regulation of apoptosis by BCL-2 family proteins. Journal of Cellular and Molecular Medicine. 2003;**7**:249-257

[5] Sato T, Iries S, Krajewski S, Reed JC. Cloning and sequencing of a DNA encoding the rat BCL-2 protein. Gene. 1994;**140**:291-292

[6] Zhong W-X, Li C, Hatzivassiliou G, Lindsten T, Yu QC, Yuan J, et al. Bax and Bak can localize to the endoplasmic reticulum to initiate apoptosis. Journal of Cell Biology. 2003;**162**:59-69

[7] Camello-Almaraz C, Gomez-Pinilla PJ, Pozo MJ, Camello PJ. Mitochondrial reactive oxygen species and Ca2+ signaling. American Journal of Physiology. Cell Physiology. 2004;**29**:C1082-C1088

[8] Yi M, Weaver D, Hajnoczky G. Control of mitochondrial motility and distribution by the calcium signal: A homeostatic circuit. Journal of Cell Biology. 2004;**167**:661-672

[9] Kroemer G, Dallaporta B, Resche-Rigon M. The mitochondrial death/life regulation in apoptosis and necrosis. Annual Review of Physiology. 1998;**60**:619-642

[10] Caputo F, Vegliante R, Ghibelli L. Redox modulation of the dna damage response. Biochemical Pharmacology. 2012;**84**:1292-1306

[11] Shareef MM, Cui N, Burikhanov R, Gupta S, Satishkumar S, Shajahan S, et al. Role of tumor factor-alpha and TRAIL in highdose radiation-induced bystander signalling in lung adenocarcinoma. Cancer Research. 2007;**67**:11811-11820

[12] Markus A, Belka C. A normal tissue dose response model of dynamic repair processes. Physics in Medicine and Biology. 2006;**51**:153-172

[13] Brenner DJ, Curtis RE, Hall EJ, Ron E. Second malignancies in prostate patients after radiotherapy compared with surgery. Cancer. 2000;**88**:398-406

[14] Pines J, Hunter T. The differential localization of human cyclins A and B is due to acytoplasmic retention signal in cyclin B. EMBO Journal. 1994;**13**:3772-3781

[15] Rodemann HP, Blaese MA. Responses of normal cells to ionizing radiation. Seminars in Radiation Oncology. 2007;**17**:8-88

[16] Ye X, Nalepa G, Welcker M, Kessler BM, Spooner E, Qin J, et al. Recognition of phosphodegron motifs in human cyclin E by the SCF (FBW7) ubiquitin ligase. Journal of Biological Chemistry. 2004;**279**:50110-50119

[17] Belka C, Jendrossek V, Pruschy M, Vink S, Verheij M, Budach W. Apoptosismodulating agents in combination with radiotherapy-current status outlook. International Journal of Radiation Oncology, Biology, Physics. 2004;**58**:542-554

[18] Orrenius S, Gogvadze V, Zhivotosky B. Mitochondrial Oxidative

Stress: Implications for cell death. Annual Review of Pharmacology and Toxicology. 2007;**47**:143-183

[19] Beutler B, Greenwald D, Hulmes JD, Chang M, Pan YC, Mathison J, et al. Identity of tumor necrosis factor and macrophage-secreted factor cachectin. Nature. 1985;**316**:552-554

[20] Claro S, Oshiro ME, Freymuller E, Katchburian E, Kallas EG, Cerri PS, et al. γ-Radiation induces apoptosis via sarcoplasmatic reticulum in Guinea pig ileum smooth muscle cells. European Journal of Pharmacology. 2008;**590**:20-28

[21] Claro S, Kanashiro CA, Oshiro ME, Ferreira AT, Khalil RA. α- and ε-protein kinase C activity during smooth muscle cell apoptosis in response to γ-radiation. Journal of Pharmacology and Experimental Therapeutics. 2007;**322**:964-972

[22] Nurse P. A long twentieth century of the cell cycle and beyond. Cell. 2000;**100**:71-78

[23] Pines J. Cyclin and their associated cyclin-dependent kinases in the human cell cycle. Biochemical Society Transactions. 1993;**21**:921-925

[24] Zhao L, Bode AM, Cao Y, Dong Z. Regulatory mechanisms and clinical perspectives of miRNA in tumor radiosensitivity. Carcinogenesis. 2012;**33**:2220-2227

[25] Attix FH. Quantities for describing the interaction of ionizing radiation with matter. In: Wiley J, Wiley S, editors. Introduction to Radiological Physics and Radiation Dosimetry. New York: Interscience; 1986. pp. 20-37

[26] Yagi K. A simple fluorometric assay for lipoperoxide in blood plasma. Biochemical Medicine. 1976;**15**:212-216

[27] Jiang Z-Y, Woollard ACS, Wolf SP. Lipid hydroperoxide measurement by oxidation of Fe++ in the presence of xylenol orange. Lipids. 1991;**26**:853-856

[28] Krishan A. Rapid flow cytofluorometric analysis of mammalian cell cycle by propidium iodide staining. The Journal of Cell Biology. 1975;**66**:188-193

[29] Kosower NS, Kosower EM. The glutathione status of cells. International Review of Cytology. 1978;**54**:109-160

[30] Hasui M, Hirabayashi Y, Kobayashi Y. Simultaneous measurement by flow cytometry of phagocytosis and hydrogen peroxide production of neutrophils in whole blood. Journal of Immunological Methods. 1989;**117**:53-58

[31] Traganos F, Darzynkiewicz Z, Melamed MR. The ratio of RNA to total nucleic acid content as a quantitative measure of unbalanced cell growth. Cytometry. 1982;**2**:212-218

[32] Claro S, Oshiro MEM, Mortara RA, Paredes-Gamero EJ, da Silva Pereira GJ, Smaili SS, et al. Radiation-generated ROS induce apoptosis via mitochondrial and cell cycle alteration in smooth muscle. International Journal of Radiation Biology. 2013

[33] Gong J, Traganos F, Darzynkiewicz Z. Growth imbalance and altered expression of cyclins B1, A, E, and D3 in MOLT-4 cells synchronized in the cell cycle by inhibitors of DNA replication. Cell Growth & Differentiation. 1995;**6**:1485-1493

[34] Johnson L, Walsh ML, Bockus BJ, Chen LB. Monitoring of relative mitochondrial membrane potential in living cells by fluorescence microscopy. The Journal of Cell Biology. 1981;**88**:526-535

[35] Hall EJ. Radiobiology for the radiologist. In: For Students of Diagnostic Radiology, Nuclear Medicine, and Radiation Oncology.

5th ed. Philadelphia: Lippincott Williams & Wilkins; 2000. pp. 51-64

[36] Edlich F, Banerjee S, Suzuki M, Cleland MM, Arnoult D, Wang C, et al. Bcl-x(L) retrotranslocates Bax from the mitochondria into the cytosol. Cell. 2011;**145**:104-116

[37] Maity A, Kao GD, Muschel RJ, McKenna WG. Potential molecular targets for manipulating the radiation response. International Journal of Radiation Oncology, Biology, Physics. 1997;**37**:639-653

[38] Halliwell B, Gutteridge JMC. Free Radicals in Biology and Medicine. 2nd ed. Oxford: Oxford Univesrsity Press; 1993. pp. 10-21, 79-80, 97-98, 196-208, 247-251

[39] Kuwana T, Mackey MR, Perkins G, Ellisman MH, Latterich M, Schneiter R, et al. Bid, Bax, and lipids cooperate to form supramolecular openings in the outer mitochondrial membrane. Cell. 2002;**111**:331-342

[40] Antonsson B, Montessuit S, Lauper S, Eskes R, Martinou JC. BAX oligomerization is required for channel-forming activity in liposomes and to trigger cytochrome c realese from mitochondria. Biochemical Journal. 2000;**345**:271-278

[41] Chipuk JE, Green DR. How do BCL-2 proteins induce mitochondrial outer membrane permeabilization? Trends in Cell Biology. 2008;**18**:157-164

[42] Kane DJ, Sarafian TA, Anton R, Hahn H, Gralla EB, Valentine JS, et al. BCL-2 inhibition of neural death: Decreased generation of reactive oxygen species. Science. 1993;**262**:1274-1277

[43] Su ZZ, Lebedeva IV, Sarkar D, Emdad L, Gupta P, Kitada S, et al. Ionizing radiation enhances therapeutic activity of mda-7/IL-24: Overcoming radiation-and mda-7/IL-24-resistance in prostate cancer cells overexpressing the antiapoptotic proteins BCL-x(L) or BCL-2. Oncogene. 2006;**25**:2339-2348

[44] Chauhan D, Hideshima T, Anderson KC. Apoptotic signalling in multiple myeloma: Therapeutic implications. International Journal of Hematology. 2003;**78**:114-120

[45] Susnow N, Zeng L, Margineantu D, Hockenbery DM. BCL-2 family proteins as regulator of oxidative stress. Seminars in Cancer Biology. 2009;**19**:42-49

[46] Liddil JD, Dorr RT, Scuderi P. Association of lysosomal activity with sensitivity and resistance to tumor necrosis factor in murine L929 cells. Cancer Research. 1989;**49**:2722-2728

[47] Waring P. Redox active calcium ion channels and cell death. Archives of Biochemistry and Biophysics. 2005;**434**:33-42

[48] Hajnoczky G, Csordas G, Das S, Garcia-Perez C, Saotome M, Sinha Roy S, et al. Mitochondrial calcium signalling and cell death: Approaches for assessing the role of mitochondrial Ca^{2+} uptake in apoptosis. Cell Calcium. 2006;**4**:553-560

[49] Lam M, Dubyak G, Chen L, Nunez G, Miesfeld RL, Distelhorst CW. Evidence that BCL-2 represses apoptosis by regulating endoplasmic reticulum-associated Ca2+ fluxes. Proceedings of the National Academy of Sciences of the United States of America. 1994;**91**:6569-6573

[50] Pizzo P, Pozzan T. Mitochondria-endoplasmic reticulum choreography: Structure and signaling dynamics. Trends in Cell Biology. 2007;**17**:511-517

[51] Scorrano L, Korsmeyer SJ. Mechanisms of cytochrome c release by proapoptotic BCL-2 family members. Biochemical and Biophysics Research Communications. 2003;**304**:437-444

[52] Touyz RM. Reactive oxygen species as mediators of calcium signaling by angiotensin implications in vascular physiology and pathophysiology. Antioxidants and Redox Signaling. 2005;**7**:1302-1304

[53] Kahl CR, Means AR. Regulation of cell cycle progression by calcium/calmodulin dependent pathways. Endocrine Reviews. 2003;**24**:719-736

[54] Choi J, Chiang A, Taulier N, Gros R, Pirani A, Husain M. A calmodulin-binding site on cyclin E mediates Ca2+–sensitive G1/S transition in vascular smooth muscle cells. Circulation Research. 2006;**98**:1273-1281

[55] Putney JW Jr. The capacitative model for receptor-activated calcium entry. In: August JT, Anders MW, Murad F, editors. Advances in Pharmacology. North Carolina: Academic Press; 1991. pp. 251-269

[56] Polavarapu R, Gongora MC, Winkles JA, Yepes M. Tumor necrosis factor-alpha weak inducer of apoptosis increases the permeability of the neurovascular unit nuclear factorkappa B pathway activation. Journal of Neuroscience. 2005;**25**:10094-100100

[57] Papakonstanti EA, Stournaras C. Tumor necrosis factor-alpha promotes survival of opossum kidney cells via Cdc42-induced phospholipase C-gamma activation and actin filament redistribution. Molecular Biology of the Cell. 2004;**15**:1273-1286

[58] Dumay A, Rincheval V, Trotot P, Mignotte B, Vayssiere JL. The superoxide dismutase inhibitor diethyldithiocarbamate has antagonistic effects on apoptosis by triggering both cytochrome c release and caspase inhibition. Free Radical Biology & Medicine. 2006;**40**:1377-1390

[59] Faleiro L, Lazebnik Y. Caspases disrupt the nuclear-cytoplasmic barrier. Journal of Cell Biology. 2000;**151**:951-959

[60] Kerr JF, Harmon BV. Definition and incidence of apoptosis: An historical perspective. In: Tomei LD, Cope FO, editors. Apoptosis: The Molecular Basis of Cell Death. Cold Spring Harbor. New York: Cold Spring Harbor Laboratory Press; 1991. pp. 5-29

[61] Cohen-Jonathan E, Bernhard EJ, McKenna WG. How does radiation kill cells? Current Opinion in Chemical Biology. 1999;**3**:77-88

[62] Livenh E, Fishman DD. Linking protein kinase C to cell-cycle control. European Journal of Biochemistry. 1997;**248**:1-9

[63] Baines CP, Song CX, Zheng YT, Wang GW, Zhang J, Wang OL, et al. Protein kinase C epsilon interacts with and inhibits the permeability transition pore in cardiac mitochondria. Circulation Research. 2003;**92**:873-880

[64] Dai Y, Pei XY, Rahmani M, Conrad DH, Dent P, Grant S. Interruption of the NFkappaB pathway by bay 11-7082 promotes UCN-01-mediated mitochondrial dysfunction and apoptosis in human myeloma cells. Blood. 2004;**103**:2761-2770

[65] Qiao LF, Xu YJ, Liu XS, Xie JG, Wang J, Du CL, et al. Role of protein kinase C alpha and cyclin D1 in the proliferation of airway smooth muscle in asthmatic rats. Chinese Medical Journal. 2008;**121**:2070-2076

[66] Kuwahara H, Nishizaki M, Kanazawa H. Nuclear localization signal and phosphorylation of Serine350 specify intracellular localization of DRAK2. Journal of Biochemistry. 2008;**143**:349-358

[67] Wu LW, Mayo LD, Dunbar JD, Kessler KM, Baerwald MR, Jaffe EA, et al. Utilization of distinct signaling

pathway by receptors for vascular endothelial cell growth factor and other mitogens in the induction of endothelial cell proliferation. Journal of Biological Chemistry. 2000;**275**:5096-5103

[68] Tachado SD, Mayhew MW, Wescott GG, Foreman TL, Goodwin CD, McJilton MA, et al. Regulation of tumor invasion and metastasis in protein kinase C epsilontransformed NIH3T3 fibroblasts. Journal of Cellular Biochemistry. 2002;**85**:785-797

[69] Takayama S, Sato T, Krajewski S, Kochel K, Irie S, Millan JA, et al. Cloning and functional analysis of BAG: A novel Bcl-2-binding protein with anti-cell death activity. Cell. 1995;**80**:279-284

[70] Takayama S, Krajewski S, Krajewska M, Kitada S, Zapata JM, Kochel K, et al. Expression and location of Hssp70/Hsc-binding anti-apoptotic protein BAG-1 and its variants in normal tissues and tumour cell lines. Cancer Research. 1998;**58**:3116-3131

[71] Gubina E, Rinaudo MS, Szallasi Z, Blumberg PM, Mufson RA. Overexpression of protein kinase C isoform epsilon but not delta in human interleukin-3-dependent cells suppresses apoptosis and induces bcl-2 expression. Blood. 1998;**91**:823-829

[72] Pardo OE, Wellbrock C, Khanzada UK, Aubert M, Arozarena I, Davidson S, et al. FGF-2 protects small cell lung cancer cells from apoptosis through a complex involving PKCeps, B-Raf and S6K2. EMBO Journal. 2006;**25**:3078-3088

Section 2

Antioxidants

Chapter 5

Effects of Oxidative Stress on Spermatozoa and Male Infertility

Yi Fang and Rongzhen Zhong

Abstract

Oxidative stress occurs when the production of potentially destructive reactive oxygen species (ROS) exceeds the body's own natural antioxidant defences, resulting in cellular damage. Spermatozoa oxidative stress is intimately linked to several reproductive pathologies including the failure of spermatozoa cryopreservation and spermatozoa-egg recognition and fertilization. In this light, this review focuses on (i) the effects of oxidative stress on spermatozoa and application of antioxidants; (ii) production of ROS during cryopreservation; and (iii) oxidative stress in male infertility. This literature describes both a physiological and a pathological role of ROS in fertility. A delicate balance between ROS necessary for physiological activity and antioxidants to protect from cellular oxidative injury is essential for fertility.

Keywords: spermatozoa, oxidative stress, antioxidants, cryopreservation, infertility

1. Effects of oxidative stress on spermatozoa

1.1 What is oxidative stress?

Oxidative stress occurs when a system has an imbalance between oxidation and reduction reactions, leading to generation of excess oxidants or molecules that accept an electron from another reactant [1]. A free radical is a molecule or element with an unpaired electron that is extremely reactive in an attempt to reach an electronically stable state. ROS are free radical derivatives of oxygen (O_2) containing molecules. Some of the clinically important ROS identified include peroxyl ($\cdot ROO^-$) and hydroxyl ($\cdot OH^-$) radicals, superoxide ($\cdot O_2^-$) anion, and H_2O_2. Nitrogen compounds such as nitric oxide (NO) and peroxynitrite anion (ONOO) also appear to play a role in oxidation and reduction reactions. Common molecules that receive the unpaired electron are lipids in membranes and carbohydrates in nucleic acids [2]. This leads to potential cellular membrane and DNA damage when ROS are greater than the antioxidant-carrying capacity (**Figure 1**).

1.2 Production of ROS

The process of mitochondrial oxidative phosphorylation uses nicotinamide adenine dinucleotide (NADH) as an electron donor and O_2 as an electron acceptor in the electron transport chain, coupling both reduction and oxidation reactions with the synthesis of adenosine triphosphate (ATP), and about 1–5% O_2 transformed into ROS [3]. Another intrinsic source of spermatic ROS production is cytoplasmic glucose-6-phosphate dehydrogenase (G-6-PDH). This cytoplasmic source of ROS

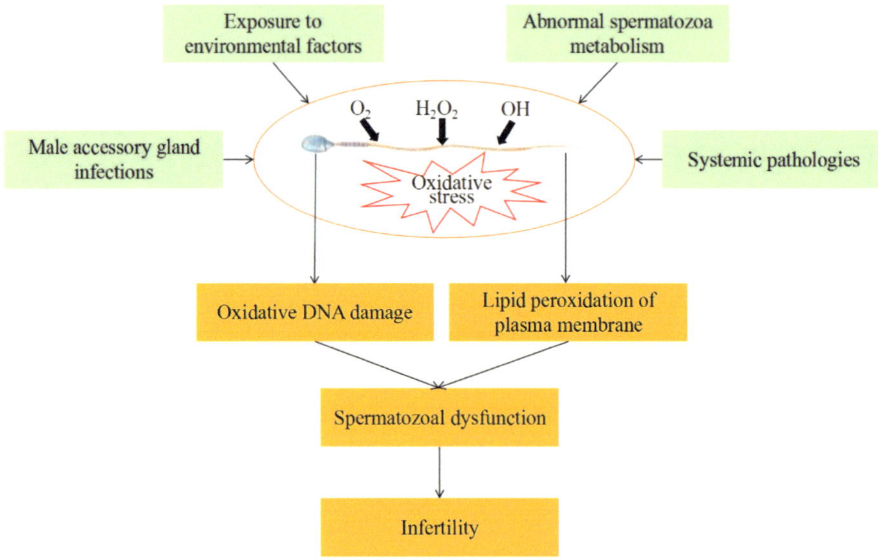

Figure 1.
Factors contributing to oxidative stress-induced male infertility.

may explain why increased spermatic cytoplasm could be linked to infertility [4]. In addition to leucocytes, infection in semen has also been implicated as a source of ROS. Exposure to heavy metals (e.g., cadmium, lead, iron and copper), pesticides, phthalate and pollution can lead to spermatozoa damage by excessive ROS [5]. Smoking has also been associated with decreased spermatic function. But industrial exposure not only induces oxidative stress but also disrupts the hypothalamic–pituitary-gonadal axis to inhibit the release of GnRH, LH and FSH in human and animal [6, 7].

1.3 Pathological effects on spermatozoa

Only the balance of ROS and antioxidants can keep the optimal spermatozoa function. Low level of ROS has been shown to be essential for fertilization, acrosome reaction, hyperactivation, motility and capacitation [8, 9]. ROS induces cyclic adenosine monophosphate (cAMP) in spermatozoa that inhibits tyrosine phosphatase, leading to tyrosine phosphorylation [10]. In particular, capacitation not only requires ROS, but also it can be inhibited by catalase (CAT) [11]. It has been described that high level of ROS can promote the acrosome reaction with the mechanism of ROS-modulated tyrosine phosphorylation [12].

1.3.1 Lipid peroxidation of plasma membrane

Lipids are present in spermatozoa plasma membrane in the form of polyunsaturated fatty acids (PUFA), most susceptible to oxidative damage [13, 14]. Once there is generation of lipid peroxide radical, it will react with the neighboring lipid molecule, triggering a chain reaction that can lead to >50% oxidation of the spermatozoa plasma membrane [15]. Byproducts of lipid oxidization include mutagenic and genotoxic molecules malondialdehyde (MDA) and 4-hydroxynonenal (4-HNE), leading indirectly to DNA damage [16]. Buffalo spermatozoa are more prone to oxidative damage than that of cattle, since it is rich in polyunsaturated fatty acids like arachidonic acids and docosahexaenoic acids [17].

1.3.2 DNA damage

Free radicals have the capability to directly damage spermatozoa DNA via single- and double-strand DNA breaks, cross-links and chromosomal rearrangements [18, 19]. ROS also can cause various types of gene mutations such as point mutations and polymorphism, resulting in decreased semen quality [20]. Other mechanisms such as denaturation and DNA base-pair oxidation also may be involved. Although most of the spermatozoa genome (85%) is bound to central nucleoprotamines that protect it from free radical attack [21], infertile men often have deficient protamination, which may make their sperm DNA more vulnerable to ROS damage [22]. A common byproduct of DNA oxidation, 8-hydroxy-2-deoxyguanosine (8-OH-2- deoxyguianosine), has been considered a key biomarker of this oxidative DNA damage [23].

1.3.3 Motility

Decreased motility has been shown to be due to ROS-induced peroxidation of lipids in the spermatozoa membrane decreasing flexibility and by inhibition of motility mechanisms [24, 25]. The axosome and associated dense fibers of the middle pieces in spermatozoa are covered by mitochondria that generate energy from intracellular stores of ATP. It is well established that ROS can induce axonemal and mitochondrial damage, resulting in the immobilization of spermatozoa [26, 27]. In addition, ROS-induced damage of mitochondrial DNA leads to decreased ATP and energy availability and leads to activation of caspases and ultimately apoptosis, impeding spermatozoa motility [28, 29]. H_2O_2 can diffuse across the membranes of spermatozoa and inhibit the activity of some vital enzymes such as glucose-6-phosphate dehydrogenase (G6PD), which is an enzyme controlling the intracellular availability of NADPH. This is used as a source of electrons by spermatozoa to fuel the generation of ROS by an enzyme system known as NADPH oxidase [30]. Another hypothesis involves a series of interrelated events resulting in a ROS-reduced motility due to a decrease in axonemal protein phosphorylation and mitochondrial membrane damage and leakage of intracellular enzymes [31]. Meanwhile, cytochrome c release during the apoptotic pathway further increases levels of ROS, promoting DNA damage and fragmentation [32]. Especially after frozen–thawed cycles, spermatozoa with higher levels of oxidative stress have higher levels of caspase activation that can trigger apoptosis [33].

1.3.4 Apoptosis

High levels of ROS disrupt the mitochondrial membranes, inducing the release of the cytochrome c protein and Ca^{2+} and activating the caspase-inducing apoptosis [34]. Apoptosis in spermatozoa also may be initiated by ROS-independent pathways involving the cell surface protein Fas, which is a type I membrane protein that belongs to the tumor necrosis factor-nerve growth factor receptor family and mediates apoptosis [35]. Mitochondrial exposure to ROS also results in the release of apoptosis-inducing factor (AIF), which directly interacts with the DNA to cause DNA fragmentation in spermatozoa [36, 37].

1.3.5 Fertilization, pregnancy and miscarriage

Lipid peroxides and DNA damage are the most typical oxidative stress injury in sperm. Lipid peroxides are spontaneously generated in the sperm plasma membrane, which induce decrease in fertility during storage of semen [38]. In addition, the importance of sperm DNA damage is brought to light when studies correlated the degree of DNA damage with various indices of fertility such as the fertilization rate, embryo cleavage rate, implantation rate, pregnancy rate and live birth rate of the offspring. If sperm DNA is unable to decondense after entering the ooplasma, fertilization may not take place or a postfertilization failure may occur when sperm DNA is defective by ROS. Higher miscarriage rate is observed with ROS-induced sperm DNA damage [39]. High-level sperm DNA fragmentation induced was related to lower pregnancy rates in in vitro fertilization (IVF) but not in intracytoplasmic sperm injection (ICSI) cycles, whereas it was associated with higher miscarriage rates in both IVF and ICSI cycles. In addition, ROS actively participate in metabolic pathways during sperm activation, which leads to cholesterol efflux, cyclic adenosine monophosphate (cAMP) production and tyrosine phosphorylation, important events that contribute to fertilization competence [40]. However, it has been also described that appropriate ROS (hydrogen peroxide stimulation) can promote the acrosome reaction and sperm hyperactivation with the mechanism of ROS-modulated tyrosine phosphorylation [41], thereby assisting the sperm's transit through the cumulus and zona pellucida [42].

2. Production of ROS during cryopreservation

Cryopreservation of spermatozoa is an applicable technique, but it may influence the post-thaw qualities of spermatozoa, including morphology, motility, viability and DNA integrity. The imbalance between the presence of ROS and spermatozoa antioxidant activity is a main cause of cryodamage of spermatozoa [43]. The specific cell structure and plasma membrane of spermatozoa, a large number of mitochondria, low cytoplasm and incomplete antioxidant system in cytoplasm make them possibly vulnerable to damage from free radicals [43]. Susceptibility to cold temperatures is also linked to a high ratio of unsaturated to saturated fatty acid content of the spermatozoa plasma membrane. Bull, ram and boar spermatozoa are more sensitive to cooling than rabbits, dogs and human, due to a higher ratio of unsaturated to saturated fatty acids [44]. Antioxidants are the main defense factors against oxidative stress induced by free radicals [45]. Supplementation of cryopreservation extenders with antioxidants provides a cryoprotective effect on bull, ram, goat, boar, canine and human spermatozoa quality, thus minimizing the detrimental effect of ROS and improving quality of post-thaw spermatozoa (**Table 1**).

Vitamin E (α-tocopherol) is a highly potent chain-breaking lipophilic antioxidant residing on the cell membrane which can break the covalent links that ROS have formed between fatty acid side chains in membrane lipids [83]. Addition of α-tocopherol in rabbit, equine, bovine, boar and ram, aiming to improve semen quality, led to inconsistent results [46]. Combined with vitamin C, vitamin E enhanced motility and viability of cooled spermatozoa [47, 48]. Askari et al. (1994) showed that vitamin E improved hypo-osmotic swelling scores and the post-thaw motility slightly. Moreover, α-tocopherol supplementation at 200 uM concentration may protect the spermatozoa against stress oxidative by reducing lipid peroxidation and DNA fragmentation [67].

The GSH content and its antioxidant defensive capacity alter during the freezing–thawing process, possibly because of oxidative stress and cell death [84], so that addition of GSH to the freezing extender has variable outcomes. Varghese et al. reported that addition of 5 mM of GSH to human spermatozoa freezing media improved the DNA integrity, but failed in reducing the lipid peroxidation and in

Antioxidant	Effects	Species	References
Vitamin C	Improve semen quality	Rabbit, equine, bovine, boar and ram	[46]
	Enhance motility and viability	Ram and goat	[47, 48]
GSH	Improve DNA integrity	Human	[49]
	Improve progressively motile, protect plasma membrane integrity	Red deer	[50]
	Ameliorate acrosome ultrastructure	Ram	[51]
	Increase motility, viability and fertilization	Boar	[52]
Vitamin C	Improve motility, acrosome and membrane integrity	Bovine	[53]
	Reduce DNA damages	Infertile human	[54]
	Reduce DNA damage and lipid peroxidation	Boar	[55]
Ergothioneine	Protect DNA integrity	Bull	[56]
Melatonin	Improve spermatozoa characteristics	Goat, rat, boar, ram, mouse and human	[57–63]
	Improve spermatozoa function	Boar	[64]
	Enhance hyperactivation	Hamster	[65]
Selenium	Ameliorate motility, viability, membrane integrity and total antioxidant capacity	Bovine	[66]
Zinc	Improve hypo-osmotic swelling (HOS), reduce lipid peroxidation and DNA fragmentation	Mammalian	[67]
Amino acids	Membrane stabilizer and inhibit spermatozoa capacitation	Ram	[68]
	Enhances membrane integrity, viability and motility, reduce lipid peroxidation and DNA damage	Bull, ram, goat, boar and fish	[69–74]
	Reduce DNA fragmentation	Fish	[75]
Natural herbs	Improve motility and viability, reduce DNA damage	Human	[76, 77]
	Enhance motility and viability and minimize DNA damage	Human, rat	[78]
	Improve viability and motility and prevent peroxidatation	Boar, canine, bull and ovine	[79–82]

Table 1.
Proposed antioxidants in spermatozoa cryopreservation.

increasing the motility [49]. Recently, Gadea et al. showed that GSH supplementation to freezing media reduced human spermatozoa ROS levels and increased the level of sulphhydryl groups on membrane proteins in spite of increasing the percentage of motile and progressively motile spermatozoa after addition of GSH to the thawing media. Longer exposure to GSH and main damaging effect on spermatozoa membrane before the dilution in the thawing extender may elucidate this difference in viability [85]. It seems that boar spermatozoa benefited from the supplementation with this antioxidant at 1 and 5 mM [86]. GSH (at 1 mM) improved the quality of red deer post-thawing spermatozoa, especially regarding kinematic parameters and mitochondrial status [50]. In ram semen, Camara et al. (2011) found no enhancement adding GSH (0.5–2 mM) to the freezing extender, but the concentration at 2 and 5 mM ameliorated the ultrastructure of the acrosome which resulted in obtaining even lower motility at 7 mM [51]. Also, adding 1–2 mM glutathione to the ram semen extender increased the activities of GPX and SOD, decreased free radicals and improved the survival rate of post-thawed spermatozoa. Addition of SOD or CAT to boar spermatozoa freezing extender not only increased spermatozoa motility and viability but also decreased post-thaw ROS generation which led to a rising in in vitro fertilizing potential of thawed spermatozoa [52]. These findings comply with results showing that the addition of CAT and SOD to the extender improved the survival and in vitro fertility of liquid stored ram spermatozoa [87].

The intake of vitamin C (ascorbic acid) could result in decreasing of GSH-Px in opposition to GSH increase and improved spermatozoa motility, acrosome and membrane integrity [53]. The addition of ascorbic acid before cryopreservation reduced DNA damages only in infertile men [54]. Because ascorbic acid is rapidly oxidized into inactive dehydroascorbate when exposed to highly oxidative environment [88], it is difficult to maintain its scavenging activities during exposure of spermatozoa to high oxidative environments for extended periods of time. Ascorbic acid 2-O-α-glucoside (AA-2G) is characterized by high resistance to thermal and oxidative degradation in neutral solutions and non-reducing conditions. Addition of AA-2G to the freezing extender improved the post-thaw quality of boar spermatozoa through the protection of spermatozoa against DNA damage and the lipid peroxidation caused by oxidative stress during cryopreservation [55].

Ergothioneine is an important low-molecular-weight thiol which scavenges singlet oxygen [89] and hydroxyl and peroxyl radicals [90]. It exists in millimolar concentrations in some tissues and has been linked to the metabolism of iron, copper and zinc. Increasing concentration of ergothioneine in semen extenders preserved DNA integrity of spermatozoa against cryodamage [56].

Melatonin (N-acetyl-5-methoxytryptamine, MT) is mainly synthesized and secreted by the pineal gland in reaction to changes in dark–light cycles [91]. It can stimulate the activity of antioxidant enzymes such as SOD and GSH-Px [92]. MT scavenges a variety of reactive oxygen and nitrogen species with powerful non-enzymatic antioxidant property [93]. MT can improve spermatozoa characteristics in goat [57], rat [58], boar [59], ram [60], mouse [61] and human [62, 63]. In addition, it had a dose-dependent effect on all parameters of spermatozoa motility. 1 uM MT did not succeed in improving the function of boar semen stored at 17°C [64], but 1 nM MT can enhance hyperactivation of hamster spermatozoa [65].

It has been indicated that dietary Se supplementation enhanced reproductive function in mice, sheep and cattle [94, 95] and also brought about the improvement in post-thaw spermatozoa quality [66]. Lack of Se has been related to reproductive problems and diminished spermatozoa quality in mice, pigs, sheep and cattle [96], but excessive Se intake also has been connected to an impaired spermatozoa quality [97]. In frozen–thawed buffalo spermatozoa, extenders containing 1 and 2 ug mL^{-1} Se significantly ameliorated spermatozoa motility, viability, membrane integrity and total antioxidant capacity. It also exerts its effects in a dose-dependent manner so that it had deleterious effects on spermatozoa parameters at high levels of 4 and 8 ug mL^{-1}.

Amino acids have an important biological role for prevention of cell damage during cryopreservation. L-cysteine (L-Cys) is a naturally occurring sulfur containing non-essential amino acid, which penetrates the spermatozoa membrane easily to participate in the intracellular GSH biosynthesis [98]. It protects the membrane lipids and proteins via indirect scavenging of free radicals; also it acts as a membrane stabilizer and inhibitor of spermatozoa capacitation [68]. Moreover, L-Cys is metabolized to taurine after passing into cells. Taurine transformed to acyl-taurine after combination with a fatty acid in plasma membrane which improves surfactant properties and osmoregulation of the spermatozoa membrane [99, 100]. It has been reported that L-Cys enhances motility and morphology of spermatozoa, reducing lipid peroxidation of plasma membrane and preventing DNA damage from ROS of post-thaw bull [69], ram [70], goat [71] and fish [72, 73] spermatozoa, and improves the viability, the chromatin structure, and membrane integrity of boar spermatozoa during chilled storage [74]; in combination with docosahexaenoic acid (DHA)-enriched hen egg yolk, L-cysteine significantly improved progressive motility and acrosome integrity of boar spermatozoa. Also, the cysteine enhanced the post-thaw Merino ram spermatozoa mitochondrial activity without improving motility after the freezing–thawing cycle. 5 or 10 mM was the optimum concentration of L-cysteine for improving the quality of frozen–thawed boar spermatozoa. Methionine had a positive effect on the sperm viability and increased the post-thaw spermatozoa motility and reduced DNA damage of fish spermatozoa [101, 102]. DNA fragmentation in gilthead seabream (*S. aurata*) and European sea bass (*D. labrax*) was significantly reduced by taurine and hypotaurine [75]. The concentration of 50 mM taurine provided the most pronounced protective effect in improving post-thaw quality of red seabream sperm [103].

The addition of natural herbs also improves the cryoprotective effect of spermatozoa. Addition of genistein to the cryoprotectant has a significant antioxidant protective effect on the frozen–thawed spermatozoa. It causes a reduction in ROS production and makes an improvement in the sperm motility and viability; it also reduces DNA damage caused by the process of cryopreservation [76, 77]. The high concentrations of genistein decreased the proportion of motile mice spermatozoa which was approved in human spermatozoa, too [104]. In ram spermatozoa, addition of either resveratrol or quercetin (5–20 ug/mL for each compound) to a Tris-egg yolk-glycerol extender decreased the mitochondrial membrane potential [105]. Quercetin at 50 uM enhanced spermatozoa motility and viability and minimized post-thawed human spermatozoa DNA damage and also proved its potential role in protecting spermatozoa against H_2O_2-mediated spermatozoa damage on spermatozoa parameters and lipid peroxidation by reducing the levels of MDA and improving activities of antioxidant enzymes in rats [78]. The antioxidant properties of *Rhodiola sacra* aqueous extract (RSAE)-enriched freezing extender with or without glycerol had substantial impacts on concentrations of MAD and GSH, apart from the quality of frozen–thawed boar spermatozoa. Likewise, the optimal concentration of RSEA in extender ranged from 4 to 8 mg L^{-1} with and without glycerol, even if the influence of 6 mg L^{-1} RSEA on spermatozoa quality was more enhanced in glycerol-free extender than glycerol-containing extender [106]. The effects of adding rosemary to semen freezing extenders in several species have been reported, including boar [79], canine [80] and ovine [81]. Rosemary-enriched freezing extender efficiently improved motility and prevented peroxidatation of epididymal boar spermatozoa, showing a significant correlation between rosemary concentration and concentration of MDA [107]. Added 10 g L^{-1} rosemary extract to the freezing extender of bull semen before cryopreservation and showed its effects on increasing viability, motility and average path velocity as well as on decreasing lipid peroxidation after thawing [82].

3. Oxidative stress in male infertility

A decline in fertility rates is becoming an increasingly prevalent issue worldwide. Infertility affects up to 15% of the population globally [108], and furthermore, male infertility is responsible in about 20% of cases but may contribute to 40% of infertile couples [109]. The leading cause of male infertility stems from a loss of spermatozoa function, ultimately resulting in a loss of fertilization potential [110]. This loss in function is causatively linked to oxidative stress within the spermatozoa driven by the presence and/or overproduction of intracellular ROS [111].

Several studies have shown conflicting results for the effect of the antioxidant therapy on male fertility, whilst a number of studies conveyed a favorable effect on basic semen parameters, advanced spermatozoa function tests and pregnancy rates. But, the ideal balance of the redox system necessary for optimal spermatozoa function is not known, and overconsumption of antioxidants may result in reductive stress that could cause detrimental effects on human health and well-being. Impairment of mitochondrial activity [112], reduction in blood–brain barrier permeability [113] and attenuation of endothelial cell proliferation [114] are consequences that have been reported to occur secondary to reductive stress. **Table 2** shows the mechanism of action of several commonly used antioxidants for the treatment of male infertility. The list of antioxidants used in treatment of male infertility is presented in **Table 3**.

Vitamin E is well accepted as the first line of defence against lipid peroxidation, protecting polyunsaturated fatty acids in cell membranes through its free radical quenching activity in biomembranes at an early state of free radical attack. MDA concentration was prevented by treatment with vitamin E; it may help in the prevention of against production of free radicals and quenches free hydroxyl radicals and superoxide anions, thereby reducing lipid peroxidation initiated by ROS at the level of plasma membranes [126]. Its antioxidant activity is similar to that of glutathione peroxidase. In infertility of male, the percentage of motile spermatozoa is significantly related to spermatozoa vitamin E content [127]. Lower levels of vitamin E were observed in the semen of infertile men [128]. Insufficient intake of vitamin E produced deleterious effects on the process of normal spermatozoa [129]. One of the earlier studies investigating vitamin E alone (300 mg daily) on infertile men reported significant improvement in spermatozoa motility [121]. Combined with clomiphene citrate treatments, vitamin E significantly improved spermatozoa concentration and motility of patients with idiopathic oligoasthenozoospermia (OAT) [130]. Another observational study investigated a daily regimen of vitamin E (400 mg) + selenium (200 lg), for a period of 100 days, on infertile men with idiopathic asthenoteratospermia. Results revealed that 52.6% of patients showed a significant improvement in spermatozoa motility, morphology or both [131]. On the other hand, a few other studies failed to reproduce any significant effect on semen

Antioxidants	Antioxidant mechanism	Typical daily dose
Vitamin E	Neutralizes free radicals	200–600 mg
Vitamin C	Neutralizes free radicals	200–1000 mg
Selenium	Enhancement of antioxidant enzyme activity	100–200 mg
Zinc	Inhibition of NADPH oxidase and scavenges hydroxyl radicals	15–40 mg
Carnitines	Neutralizes free radicals and acts as an energy source	1–3 g
CoQ10	Scavenges free radicals of mitochondrial electron transport system	60–300 mg
Lycopene	Scavenges free radicals	4–6 mg

Vitamin E, tocopherol; vitamin C, ascorbic acid; NADPH, nicotinamide adenine dinucleotide phosphate.

Table 2.
Mechanism of action of commonly used antioxidants and clinical dosage.

Clinical applications	Antioxidants daily	References
Oligozoospermia	Vitamin E (180 mg), vitamin A (30 mg) and essential fatty acids (600 mg)	[22]
	LC (2 g)	[115]
	CoQ10 (300 mg)	[116]
	Selenium (200 mg)	[117]
	Folic acid (5 mg) + zinc (66 mg)	[118]
	Lycopene (2 mg)	[119]
Asthenozoospermia	Zinc (400 mg), vitamin E (20 mg) and vitamin C (10 mg)	[120]
	CoQ10 (300 mg)	[116]
	Selenium (200 mg)	[117]
	Lycopene (2 mg)	[119]
Teratozoospermia	Selenium (200 mg)	[117]
	Zinc (400 mg), vitamin E (20 mg) and vitamin C (10 mg)	[120]
Improving DNA integrity	Vitamin E (1 g) + vitamin C (1 g)	[121]
	Vitamin C (400 mg), vitamin E (400 mg), b-carotene (18 mg), zinc (500 mmol) and selenium (1 mmol)	[29]
	LC (1500 mg), vitamin C (60 mg), CoQ10 (20 mg), vitamin E (10 mg), zinc (10 mg), folic acid (200 lg),	[32]
Improving ART	vitamin E (200 mg)	[33]
	Vitamin E (600 mg)	[122]
	Vitamin C (1 g) + vitamin E (1 g)	[123]
Improving live birth rate	CoQ10 (300 mg)	[116]
	Vitamin E (300 mg)	[124]
	Zinc (5000 mg)	[12]
	Vitamin E (1 g) + vitamin C (1 g)	[121]
	Carnitines: LC (2 g) + LAC (1 g)	[125]

Table 3.
Proposed antioxidants in various clinical treatments.

parameters using vitamin E as a single treatment [123, 125] or in combination with other antioxidants [132].

In the male reproductive system, vitamin C (ascorbic acid) is known to protect spermatogenesis and plays a key role in spermatozoa integrity and fertility both in men by increasing testosterone levels and preventing spermatozoa agglutination. It exists at a concentration 10 times higher in seminal plasma than in blood serum [133] and contributes up to 65% of the total antioxidant capacity of seminal plasma found intracellularly and extracellularly [134, 135]. Semen of infertile men with asthenozoospermia was found to contain lower vitamin C levels and higher ROS levels than those obtained from fertile controls [117]. Vitamin C as a single agent which is used to treat heavy smokers, with a daily dose of 200 or 1000 mg or placebo for 1 month, significantly improved spermatozoa quality [136]. Receiving 500 mg daily vitamin C with a combination of zinc, vitamin E and vitamin C for a total of 3 months after undergoing varicocelectomy significantly improved spermatozoa motility and morphology on varicocelectomy patients [118, 137].

Carnitines [L-carnitine (LC) and L-acetyl carnitine (LAC)] are water-soluble antioxidants involved in spermatozoa metabolism, fuelling important activities like spermatozoa motility [138]. The carnitine and acetylcarnitine can significantly improve spermatozoa motility or kinetics in patients with asthenozoospermia [120, 139]. In vitro studies of spermatozoa cultured in media containing carnitines had higher motility and viability. They exhibit their antioxidant activities through scavenging superoxide anions and hydrogen peroxide radicals, thereby inhibiting lipid peroxidation. A combined treatment of LC (2 g) and LAC (1 g) for 2 months' duration to placebo in men with OAT showed significant improvement in all semen parameters; however, the most significant increase was in spermatozoa motility. Low-grade varicocele and idiopathic infertility patients treated with LC and LAC in comparison with placebo had significant improvement in all semen parameters [140]. On the contrary, LC (1000 mg) and LAC (500 mg) daily treated asthenozoospermic men for 12 weeks and failed to show any significant improvement in spermatozoa motility [141].

CoQ10 is a vital antioxidant omnipresent in almost all body tissues. It is particularly present at high concentrations in spermatozoa mitochondria involved in cellular respiration and plays an integral role in energy production [142]. This contribution rationalizes its use as a promotility and antioxidant molecule. Furthermore, CoQ10 inhibits superoxide formation, delivering protection against OS-induced spermatozoa dysfunction. A significant negative correlation between CoQ10 levels and hydrogen peroxide has been reported, and a linear correlation between CoQ10 levels in seminal plasma spermatozoa count and motility was detected [115]. 300 mg CoQ10 for 26 weeks obtained a significant increase in sperm density and motility [143]. A systemic review of clinical trials on 332 infertile men revealed that treatment with CoQ10 (200–300 mg daily) resulted in a significant increase in spermatozoa concentration and motility [144].

Antioxidant properties of selenium are thought to stem from its ability to augment the function of glutathione. More than 25 selenoproteins exist, such as phospholipid hydroperoxide glutathione peroxidase (PHGPX) [145] and spermatozoa capsular selenoprotein glutathione peroxidase [146], to maintain spermatozoa structural integrity [147]. Selenium deficiency has been most commonly associated with morphological spermatozoa midpiece abnormalities and impairment of spermatozoa motility [148]. A significant increase (74%) in total normal spermatozoa concentration was noted amongst the subfertile group receiving combined therapy [116] with a combination of both folic acid and zinc for 26 weeks of treatment. Selenium has been less frequently investigated for the treatment of subfertile men. As previously noted, with selenium (200 mg) supplements for 26 weeks, results showed a significant improvement in all semen parameters. A strong correlation was seen between the sum of the selenium and mean spermatozoa concentration, motility and percentage normal morphology [149]. Furthermore, the combination of selenium with vitamin E resulted in an increase in spermatozoa motility [124, 150]. But in the contrary report, treatment with selenium (300 mg) daily for 48 weeks did not result in a significant influence on semen parameters of a group of normozoospermic men [122].

Zinc plays a vital role in the metabolism of RNA and DNA, signal transduction, gene expression and regulation of apoptosis. Its antioxidant properties are thought to result from its ability to decrease production of hydrogen peroxide and hydroxyl radicals through antagonizing redox-active transition metals, such as iron and copper [151]. Zinc concentrations of seminal plasma were found to be significantly lower in subfertile men [152]. Spermatozoa flagellar abnormalities, such as hypertrophy and hyperplasia of the fibrous sheath, axonemal disruption, defects of the inner microtubular dynein arms and abnormal or absent midpiece, are all associated with zinc deficiency [153]. Zinc given for 3 months in men with asthenozoospermia obtained a significant improvement in spermatozoa concentration, progressive motility and fertilizing capacity and a reduction in the incidence of anti-spermatozoa antibodies [153]. Oral zinc supplementation successfully restored seminal catalase-like activity and improved spermatozoa concentration and progressive motility in a group of asthenozoospermic men [154].

Lycopene is a naturally synthesized carotenoid presented in fruits and vegetables. Its powerful ROS quenching abilities make it a major contributor to the human redox defense system [155]. Lycopene is detected at high concentrations in human testes and seminal plasma with levels that tend to be lower in infertile men [156]. The treatment with 2 mg lycopene twice daily for 3 months significantly improves spermatozoa concentration and motility in 66% of patients, respectively. However, the effects were only significant in patients who had baseline spermatozoa concentrations of $>5 \times 10^6$ sperm/mL [119].

4. Conclusion

Spermatozoa possess an inherent but limited capacity to generate ROS which may help the fertilization process. Antioxidants improve the motility and fertilizing ability of spermatozoa. A balance between the benefits and risks from ROS and antioxidants appears to be necessary for the survival and normal functioning of spermatozoa. Antioxidants in extenders may minimize the detrimental effect of ROS and improve the quality of frozen–thawed spermatozoa in animals and human. From the other point of view, the divergent effect of each antioxidant supplementation, improving different parameters of frozen–thawed sperm quality, is attributed to animal species, extender medium and type of molecule and concentration used for each species. Although a beneficial influence was generally observed for antioxidants in reversing ROS-induced spermatozoa dysfunction and in improving pregnancy rates, evaluation of ROS and the use of antioxidants are not routine in clinical practice. The dose and duration of these antioxidants should also be determined and standardized. There should be an effort to develop optimum combinations of antioxidants to supplement spermatozoa media. Finally, this study suggests that further research should be done to determine the appropriate antioxidant compounds as well as certain dose of antioxidants whether used clinical practices or cryopreservation. Moreover the future studies should concern the spermatozoa fertilization and pregnancy rate as a research emphasis.

Author details

Yi Fang[1,2]* and Rongzhen Zhong[1,2]

1 Grassland Agri-Husbandry Research Center, College of Grassland Science, Qingdao Agricultural University, Qingdao, China

2 Jilin Provincial Key Laboratory of Grassland Farming, Northeast Institute of Geography and Agroecology, Chinese Academy of Sciences, Changchun, Jilin, China

*Address all correspondence to: fangyi@iga.ac.cn

References

[1] Henkel RR. Leukocytes and oxidative stress: Dilemma for sperm function and male fertility. Asian Journal of Andrology. 2011;**13**:43-52

[2] Tremellen K. Oxidative stress and male infertility a clinical perspective. Human Reproduction Update. 2008;**14**:243-258

[3] Boveris A, Chance B. The mitochondrial generation of hydrogen peroxide. General properties and effect of hyperbaric oxygen. The Biochemical Journal. 1973;**134**:707-716

[4] Gomez E, Buckingham DW, Brindle J, Lanzafame F, Irvine DS, Aitken RJ. Development of an image analysis system to monitor the retention of residual cytoplasm by human spermatozoa: Correlation with biochemical markers of the cytoplasmic space, oxidative stress, and sperm function. Journal of Andrology. 1996;**17**:276-287

[5] Gharagozloo P. Treating Male Infertility Secondary to Sperm Oxidative Stress. US Patent Application. 2015;**15**:939-953

[6] Queiroz EK, Waissmann W. Occupational exposure and effects on the male reproductive system. Cadernos de Saúde Pública. 2006;**22**:485-493

[7] Ni K, Steger K, Yang H, Wang H, Hu K, Zhang T. A comprehensive investigation of sperm DNA damage and oxidative stress injury in infertile patients with subclinical, normozoospermic, and astheno/oligozoospermic clinical varicocele. Andrology. 2016;**4**:816-824

[8] Agarwal A, Nallella KP, Allamaneni SS, Said TM. Role of antioxidants in treatment of male infertility: An overview of the literature. Reprod. Biomed. 2004;**8**:616-627

[9] Gagnon C, Iwasaki A, De Lamirande E, Kovalski N. Reactive oxygen species and human spermatozoa. Annals of the New York Academy of Sciences. 1991;**637**:436-444

[10] Tremellen K, Miari G, Froiland D, Thompson J. A randomised control trial examining the effect of an antioxidant (Menevit) on pregnancy outcome during IVF-ICSI treatment. The Australian and New Zealand Journal of Obstetrics and Gynaecology. 2007;**47**:216-221

[11] Omu AE, Dashti H, Al-Othman S. Treatment of asthenozoospermia with zinc sulphate: Andrological, immunological and obstetric outcome. European Journal of Obstetrics, Gynecology, and Reproductive Biology. 1998;**79**:179-184

[12] Cavallini G, Ferraretti AP, Gianaroli L, Biagiotti G, Vitali G. Cinnoxicam and L -carnitine/acetyl-L-carnitine treatment for idiopathic and varicocele-associated oligoasthenospermia. Journal of Andrology. 2004;**25**:761-772

[13] Kodama H, Kuribayashi Y, Gagnon C. Effect of sperm lipid peroxidation on fertilization. Journal of Andrology. 1996;**17**:151-157

[14] Griveau JF, Dumont E, Renard P, Callegari JP, Lannou DL. Reactive oxygen species, lipid peroxidation and enzymatic defence systems in human spermatozoa. Journal of Reproduction and Fertility. 1995;**103**:17-26

[15] Sigman M, Glass S, Campagnone J, Pryor JL. Carnitine for the treatment of idiopathic asthenospermia: A randomized, double-blind, placebo-controlled trial. Fertility and Sterility. 2006;**85**:1409-1414

[16] Huang XX, Zhang AH, Hong F. Expression of MDA and 4-HNE AFB 1-induced rat hepatic putative preneoplastic lesions alleviated by Li_2CO_3 and their significance. Carcinogenesis, Teratogenesis and Mutagenesis. 2005;**3**:222-228

[17] Singh P, Chand D, Georgie GC. Effect of vitamin E on lipid peroxidation in buffalo *Bubalus bubalis*. Indian Journal of Experimental Biology. 1989;**27**:14-16

[18] Kemal DN, Morshedi M, Oehninger S. Effects of hydrogen peroxide on DNA and plasma membrane integrity of human spermatozoa. Fertility and Sterility. 2000;**74**:1200-1207

[19] Aitken RJ, Krausz C. Oxidative stress, DNA damage and the Y chromosome. Reproducción. 2001;**122**:497-506

[20] Spiropoulos J, Turnbull DM, Chinnery PF. Can mitochondrial DNA mutations cause sperm dysfunction? Molecular Human Reproduction. 2002;**8**:719-721

[21] Ammar O, Haouas Z, Hamoud B, Hamdi H, Hellara I, Jlali A, et al. Relationship between sperm DNA damage with sperm parameters, oxidative markers in teratozoospermic men. European Journal of Obstetrics and Gynecology and Reproductive Biology. 2019;**233**:70-75

[22] Gupta NP, Kumar R. Lycopene therapy in idiopathic male infertility—A preliminary report. International Urology and Nephrology. 2002;**34**:369-372

[23] Helbock HJ, Beckman KB, Shigenaga MK, Walter PB, Woodall AA, Yeo HC. DNA oxidation matters: The HPLC-electrochemical detection assay of 8-oxo-deoxyguanosine and 8-oxo-guanine. Proceedings of the National Academy of Sciences of the United States of America. 1998;**95**:93-95

[24] Lenzi A, Lombardo F, Gandini L, Alfano P, Dondero F. Computer assisted sperm motility analysis at the moment of induced pregnancy during gonadotropin treatment for hypogonadotropic hypogonadism. Journal of Endocrinological Investigation. 1993;**16**:683-686

[25] Armstrong JS, Rajasekaran M, Chamulitrat W, Gatti P, Hellstrom WJ, Sikka SC. Characterization of reactive oxygen species induced effects on human spermatozoa movement and energy metabolism. Free Radical Biology and Medicine. 1999;**26**:869-880

[26] Aitken RJ, Clarkson JS. Cellular basis of defective sperm function and its association with the genesis of reactive oxygen species by human spermatozoa. Journal of Reproduction and Fertility. 1987;**81**:459-469

[27] Peris IS, Bilodeau JF, Dufour M, Bailey J. Impact of cryopreservation and reactive oxygen species on DNA integrity, lipid peroxidation, and functional parameters in ram semen. Molecular Reproduction and Development. 2007;**74**:878-892

[28] Menezo YJ, Hazout A, Panteix G, Robert F, Rollet J, Cohen-Bacrie P. Antioxidants to reduce sperm DNA fragmentation: An unexpected adverse effect. Reproductive Biomedicine Online. 2007;**14**:418-421

[29] Gual-Frau J, Abad C, Amengual MJ, Hannaoui N, Checa MA, Ribas-Maynou J. Oral antioxidant treatment partly improves integrity of human sperm DNA in infertile grade I varicocele patients. Human Fertility (Cambridge, England). 2015;**18**:225-229

[30] Aitken RJ. Molecular mechanisms regulating human sperm function. Molecular Human Reproduction. 1997;**3**:169-173

[31] de Lamirande E, Gagnon C. Reactive oxygen species and human spermatozoa. II. Depletion of adenosine triphosphate plays an important role in the inhibition of sperm motility. Journal of Andrology. 1992;**13**:379-386

[32] Geva E, Bartoov B, Zabludovsky N, Lessing JB, Lerner-Geva L, Amit A. The effect of antioxidant treatment on human spermatozoa and fertilization rate in an in vitro fertilization program. Fertility and Sterility. 1996;**66**:430-434

[33] Safarinejad MR. The effect of coenzyme Q(10) supplementation on partner pregnancy rate in infertile men with idiopathic oligoasthenoteratozoospermia: An open-label prospective study. International Urology and Nephrology. 2012;**44**:689-700

[34] Lee E, Ahn MY, Kim HJ, Kim IY, Han SY, Kang TS, et al. Effect of di(n-butyl) phthalate on testicular oxidative damage and antioxidant enzymes in hyperthyroid rats. Environmental Toxicology. 2007;**22**:245-255

[35] Krammer PH, Behrmann I, Daniel P, Dhein J, Debatin KM. Regulation of apoptosis in the immune system. Current Opinion in Immunology. 1994;**6**:279-289

[36] Paasch U, Sharma RK, Gupta AK, Grunewald S, Mascha EJ, Thomas JAJ. Cryopreservation and thawing is associated with varying extent of activation of apoptotic machinery in subsets of ejaculated human spermatozoa. Biology of Reproduction. 2004;**71**:1828-1837

[37] Cande C, Cecconi F, Dessen P, Kroemer G. Apoptosis-inducing factor (AIF): Key to the conserved caspase-independent pathways of cell death? The Journal of Cell Science. 2002;**115**:4727-4734

[38] Twigg J, Fulton N, Gomez E, Irvine DS, Aitken RJ. Analysis of the impact of intracellular reactive oxygen species generation on the structural and functional integrity of human spermatozoa: Lipid peroxidation, DNA fragmentation and effectiveness of antioxidants. Human Reproduction. 1998;**13**:1429-1436

[39] Carrell D, Liu L, Peterson C, Jones K, Hatasaka H, Erickson L. Sperm DNA fragmentation is increased in couples with unexplained recurrent pregnancy loss. Archives of Andrology. 2003;**49**:49-55

[40] Aitken RJ, Nixon B. Sperm capacitation: A distant landscape glimpsed but unexplored. Molecular Human Reproduction. 2013;**19**:785-793

[41] Zhao J, Zhang Q, Wang YG, Li YP. Whether sperm deoxyribonucleic acid fragmentation has an effect on pregnancy and miscarriage after in vitro fertilization/intracytoplasmic sperm injection: A systematic review and meta-analysis. Fertility and Sterility. 2014;**102**:998-1005

[42] de Lamirande E, Gagnon C. A positive role for the superoxide anion in triggering hyperactivation and capacitation of human spermatozoa. International Journal of Andrology. 1993;**16**:21-25

[43] Slaweta R, Wasowicz W, Laskowska T. Selenium content, glutathione peroxidase activity, and lipid peroxide level in fresh bull semen and its relationship to motility of spermatozoa after freezing-thawing. Zentralblatt für Veterinärmedizin. Reihe A. 1988;**35**:455-460

[44] Noiles EE, Bailey J, Storey BT. Temperature dependence of the water permeability, Lp, of murine sperm shows a discontinuity between 4 and 0°C. Cryobiology. 1995;**32**:220-238

[45] Upreti GC, Jensen K, Munday R, Duganzich DM, Vishwanath R,

Smith JF. Studies on aromatic amino acid oxidase activity in ram spermatozoa: Role of pyruvate as an antioxidant. Animal Reproduction Science. 1998;**51**:275-287

[46] Hatamoto LK, Sobrinho CB, Nichi M, Barnabe V, Barnabe R, Cortada C. Effects of dexamethasone treatment (to mimic stress) and Vitamin E oral supplementation on the spermiogram and on seminal plasma spontaneous lipid peroxidation and antioxidant enzyme activities in dogs. Theriogenology. 2006;**66**:1610-1614

[47] Benhenia K, Rahab H, Smadi MA, TakfarinasIdres HB, Ouada MI. Beneficial and harmful effects of cyclodextrin-vitamin E complex on cryopreserved ram sperm. Animal Reproduction Science. 2018;**195**:266-273

[48] Amidi F, Farshad A, Khor AK. Effects of cholesterol-loaded cyclodextrin during freezing step of cryopreservation with TCGY extender containing bovine serum albumin on quality of goat spermatozoa. Cryobiology. 2010;**61**:94-99

[49] Varghese A, Das S, Bhattacharya A, Bhattacharya S, Mandal M, Agarwal A. Effect of cryoprotective additives-reduced glutathione, acetyl-L-carnitine on sperm membrane lipid peroxidation, DNA integrity and recovery of motile human sperm. Fertility and Sterility. 2005;**84**:410-411

[50] Luis AL, Manuel AR, Olga GA, Mercedes A, Maroto-Morales A, Anel L, et al. Reduced glutathione and Trolox (vitamin E) as extender supplements in cryopreservation of red deer epididymal spermatozoa. Animal Reproduction Science. 2012;**135**:37-46

[51] Silva S, Soares A, Batista A, Almeida F, Nunes J, Peixoto C, et al. In vitro and in vivo evaluation of ram sperm frozen in tris egg-yolk and supplemented with superoxide dismutase and reduced glutathione. Reproduction in Domestic Animals. 2011;**46**:874-881

[52] Roca J, Carvajal G, Lucas X, Vazquez JM, Martinez EA. Fertility of weaned sows after deep intrauterine insemination with a reduced number of frozen-thawed spermatozoa. Theriogenology. 2003;**60**:77-87

[53] Hu JH, Tian WQ, Zhao XL, Zan LS, Wang H, Li QW, et al. The cryoprotective effects of ascorbic acid supplementation on bovine semen quality. Animal Reproduction Science. 2010;**121**:72-77

[54] Branco CS, Garcez ME, Pasqualotto FF, Erdtman B, Salvador M. Resveratrol and ascorbic acid prevent DNA damage induced by cryopreservation in human semen. Cryobiology. 2010;**60**:235-237

[55] Yoshimoto T, Nakamura S, Yamauchi S, Muto N, Nakada T, Ashizawa K, et al. Improvement of the postthaw qualities of Okinawan native pig spermatozoa frozen in an extender supplemented with ascorbic acid 2-O-α-glucoside. Cryobiology. 2008;**57**:30-36

[56] Motohashi N, Mori I, Sugiura Y. Complexing of copperion by ergothioneine. Chemical and Pharmaceutical Bulletin. 1976;**24**:2364

[57] Ramadan T, Taha T, Samak M, Hassan A. Effectiveness of exposure to longday followed by melatonin treatment on semen characteristics of Damascus male goats during breeding and non-breeding seasons. Theriogenology. 2009;**71**:458-468

[58] Sonmez M, Yuce A, Turk G. The protective effects of melatonin and vitamin E on antioxidant enzyme activities and epididymal sperm characteristics of homocysteine treated

male rats. Reproductive Toxicology. 2007;**23**:226-231

[59] Lane RL, Whitaker BD. Melatonin and tannic acid supplementation in vitro improve fertilization and embryonic development in pigs. Animal Reproduction. 2018;**15**:118-123

[60] Ashrafi I, Kohram H, Naijian H, Bahreini M, Poorhamdollah M. Protective effect of melatonin on sperm motility parameters on liquid storage of ram semen at 5°C. African Journal of Biotechnology. 2011;**10**:66-70

[61] Sarabia L, Maurer I, Bustos-Obregon E. Melatonin prevents damage elicited by the organophosphorous pesticide diazinon on mouse sperm DNA. Ecotoxicology and Environmental Safety. 2009;**72**:663-668

[62] Espino J, Bejarano I, Ortiz A, Lozano GM, Garca JF, Pariente JA, et al. Melatonin as a potential tool against oxidative damage and apoptosis in ejaculated human spermatozoa. Fertility and Sterility. 2010;**94**:1915-1917

[63] Ortiz A, Espino J, Bejarano I, Lozano GM, Monllor F, Garcia JF, et al. Highendogenousmelatoninconcentrations enhance sperm quality and short-term in vitro exposure to melatonin improves aspects of sperm motility. Journal of Pineal Research. 2011;**50**:132-139

[64] Martin-Hidalgo D, Baron F, Bragado M, Carmona P, Robina A, Garcia-Marin L, et al. The effect of melatonin on the quality of extended boar semen after long-term storage at 17°C. Theriogenology. 2011;**75**:1550-1560

[65] Fujinoki M. Melatonin-enhanced hyperactivation of hamster sperm. Reproduction. 2008;**136**:533-541

[66] Abdel-Halim BR, Helmy NA. Effect of nano-selenium and nano-zinc particles during in vitro maturation on the developmental competence of bovine oocytes. Animal Production Science. 2018;**58**:2021-2028

[67] Karl K, Michal Z, Erma Z, Drobnis MS, Sutovsky P. Zinc ion flux during mammalian sperm capacitation. Nature Communications. 2018;**9**:206

[68] Coyan K, Baspınar N, Bucak MN, Akalin PP. Effects of cysteine and ergothioneine on post-thawed merino ram sperm and biochemical parameters. Cryobiology. 2011;**63**:1-6

[69] Bilodeau JF, Blanchette S, Gagnon C, Sirard MA. Thiols prevent H_2O_2-mediated loss of sperm motility in cryopreserved bull semen. Theriogenology. 2001;**56**:275-286

[70] Andreea A, Stela Z. Role of antioxidant additives in the protection of the cryopreserved semen against free radicals. Romanian Biotechnology Letters. 2010;**15**:33-41

[71] Bucak MN, Uysal O. The role of antioxidants in freezing of Saanen goat semen. The Indian Veterinary Journal. 2008;**85**:148-150

[72] Szczesniak-Fabianczyk B, Bochenek M, Smorag Z, Ryszka F. Effect of antioxidants added to boar semen extender on the semen survival time and sperm chromatin structure. Reproductive Biology. 2003;**3**:81-87

[73] Stejskal K, Svobodova Z, Fabrik I, Adam V, Beklova M, Rodina M, et al. Content of cysteine, reduced and oxidized glutathione in spermatozoa of representatives of Acipenseriformes (*Acipenser baerii* and *A. ruthenus*) as well as teleosts (*Perca fluviatilis* and *Sander lucioperca*). Journal of Applied Ichthyology. 2008;**24**:519-521

[74] Ogretmen F, Inanan BE, Kutluyer F, Kaym M. Effect of semen extender supplementation with cysteine on post-thaw sperm quality, DNA damage,

and fertilizing ability in the common carp (*Cyprinus carpio*). Theriogenology. 2015;**83**:1548-1552

[75] Cabrita E, Ma S, Diogo P, Martínez-Paramo S, Sarasquete C, Dinis MT. The influence of certain amino acids and vitamins on post-thaw fish sperm motility, viability and DNA fragmentation. Animal Reproduction Science. 2011;**125**:189-195

[76] Thomson L, Fleming S, Aitken R, De Iuliis G, Zieschang JA, Clark A. Cryopreservation-induced human sperm DNA damage is predominantly mediated by oxidative stress rather than apoptosis. Human Reproduction. 2009;**24**:2061-2070

[77] Sierens J, Hartley J, Campbell M, Leathem A, Woodside J. In vitro isoflavone supplementation reduces hydrogen peroxide-induced DNA damage in sperm. Teratogenesis, Carcinogenesis, and Mutagenesis. 2002;**22**:227-234

[78] Abdallah FB, Zribi N, Ammar-Keskes L. Antioxidative potential of Quercetin against hydrogen peroxide induced oxidative stress in spermatozoa in vitro. Andrologia. 2011;**43**:261-265

[79] Malo C, Gil L, Gonzalez N, MartInez F, Cano R, De Blas I, et al. Anti-oxidant supplementation improves boar sperm characteristics and fertility after cryopreservation: Comparison between cysteine and rosemary (*Rosmarinus officinalis*). Cryobiology. 2010;**61**:142-147

[80] Gonzalez N, Gil L, Martinez F, Malo C, Cano R, Mur P, et al. Effect of natural antioxidant rosemary in canine soya freezing extender. Reproduction in Domestic Animals. 2010;**45**:88

[81] Gil L, Mascaro F, Mur P, Gale I, Silva A, Gonzalez N. Freezing ram semen: The effect of combination of soya and rosemary essences as a freezing extender on post-thaw sperm motility. Reproduction in Domestic Animals. 2010;**45**:91

[82] Daghigh-Kia H, Olfati-Karaji R, Hoseinkhani A, Ashrafi I. Effect of rosemary (*Rosmarinus officinalis*) extracts and glutathione antioxidants on bull semen quality after cryopreservation. Spanish Journal of Agricultural Research. 2014;**12**:98-105

[83] Jeong YJ, Kim MK, Song HJ, Kang EJ, Ock SA, Kumar BM, et al. Effect of α-tocopherol supplementation during boar semen cryopreservation on sperm characteristics and expression of apoptosis related genes. Cryobiology. 2009;**58**:181-189

[84] Bilodeau JF, Chatterjee S, Sirard MA, Gagnon C. Levels of antioxidant defenses are decreased in bovine spermatozoa after a cycle of freezing and thawing. Molecular Reproduction and Development. 2000;**55**:282-288

[85] Gadea J, Molla M, Selles E, Marco M, Garcia-Vazquez F, Gardon J. Reduced glutathione content in human sperm is decreased after cryopreservation: Effect of the addition of reduced glutathione to the freezing and thawing extenders. Cryobiology. 2011;**62**:40-46

[86] Gadea J, Garcıa-Vazquez F, Matas C, Gardon JC, Canovas S, Gumbao D. Cooling and freezing of boar spermatozoa: Supplementation of the freezing media with reduced glutathione preserves sperm function. Journal of Andrology. 2005;**26**:396-404

[87] Maxwell W, Stojanov T. Liquid storage of ram semen in the absence or presence of some antioxidants. Reproduction, Fertility, and Development. 1996;**8**:1013-1020

[88] Linster CL, Van Schaftingen E, Vitamin C. Biosynthesis, recycling and

degradation in mammals. The FEBS Journal. 2007;**274**:1-22

[89] Dahl TA, Midden WR, Hartman PE. Some prevalent biomolecules as defenses against singlet oxygen damage. Photochemistry and Photobiology. 1988;**47**:357-362

[90] Asmus KD, Benasson RV, Bernier JL, Houssin R. One electron oxidation of ergothioneine and analogues investigated by pulse radiolysis: Redox reaction involving ergothioneine and vitamin C. The Biochemical Journal. 1996;**315**:625-629

[91] Panke ES, Rollag MD, Reitter RJ. Pineal melatonin concentrations in the Syrian hamster. Endocrinology. 1979;**104**:194-197

[92] Jang H, Kim YH, Kim BW. Ameliorative effects of melatonin against hydrogen peroxide-induced oxidative stress on boar sperm characteristics and subsequent in vitro embryo development. Reproduction in Domestic Animals. 2010;**45**:943-950

[93] Raygan F, Ostadmohammadi V, Bahmani F, Reiter RJ, Asemi Z. Melatonin administration lowers biomarkers of oxidative stress and cardio-metabolic risk in type 2 diabetic patients with coronary heart disease: A randomized, double-blind, placebo-controlled trial. Clinical Nutrition. 2019;**38**:191-196

[94] Chiachun T, Hong C, Haifun R. The effects of selenium on gestation, fertility, and offspring in mice. Biological Trace Element Research. 1991;**30**:227-231

[95] Sanders D. Use of selenium in problem cattle herds. Modern Veterinary Practice. 1984;**65**:136-138

[96] Wagner H, Cheng JW, Ko EY. Role of reactive oxygen species in male infertility: An updated review of literature. Arab Journal of Urology. 2018;**16**:35-43

[97] Jamali NU, Kaka A, Khatri P, Malhi M, Naeem M, Memon AA, et al. Effect of in vitro selenium addition to the semen extender on the spermatozoa characteristics before and after freezing in Kundhi Buffalo Bull and in vivo fertility rate. Pakistan Journal of Zoology. 2019;**51**:317-323

[98] Uysal O, Bucak M. Effects of oxidized glutathione, bovine serum albumin, cysteine and lycopene on the quality of frozen-thawed ram semen. Acta Veterinaria. 2007;**76**:383-390

[99] Amidi F, Pazhohan A, Nashtaei MS, Khodarahmian M, Nekoonam S. The role of antioxidants in sperm freezing: A review. Cell and Tissue Banking. 2016;**17**:745-756

[100] Esteves SC, Sharma RK, Thomas AJ, Agarwal A. Evaluation of acrosomal status and sperm viability in fresh and cryopreserved specimens by the use of fluorescent peanut agglutinin lectin in conjunction with hypo-osmotic swelling test. International Brazilian Journal of Urology. 2007;**33**:364-376

[101] Lahnsteiner F. The role of free amino acids in semen of rainbow trout *Oncorhynchus mykiss* and carp *Cyprinus carpio*. Journal of Fish Biology. 2009;**75**:816-833

[102] Kutluyer F, Ogretmen F, Inanan BE. Effects of semen extender supplemented with L-methionine and packaging methods (straws and pellets) on post-thaw goldfish (*Carassius auratus*) sperm quality and DNA damage. CryoLetters. 2015;**36**:336-343

[103] Liu Q, Wang X, Wang W, Zhang X, Xu S, Ma D. Effect of the addition of six antioxidants on sperm motility, membrane integrity and mitochondrial function in red seabream (*Pagrus major*) sperm cryopreservation.

Fish Physiology and Biochemistry. 2014;**41**:413-422

[104] Bajpai M, Doncel G. Involvement of tyrosine kinase and cAMP-dependent kinase cross-talk in the regulation of human sperm motility. Reproduction. 2003;**126**:183-195

[105] Silva E, Cajueiro J, Silva S, Soares P, Guerra M. Effect of antioxidants resveratrol and quercetin on in vitro evaluation of frozen ram sperm. Theriogenology. 2012;**77**:1722-1726

[106] Zhao HW, Li QW, Ning GZ, Han ZS, Jiang ZL, Duan YF. Rhodiola sacra aqueous extract (RSAE) improves biochemical and sperm characteristics in cryopreserved boar semen. Theriogenology. 2009;**71**:849-857

[107] Malo C, Gil L, Cano R, MartInez F, Gale I. Antioxidant effect of rosemary (*Rosmarinus officinalis*) on boar epididymal spermatozoa during cryopreservation. Theriogenology. 2011;**75**:1735-1741

[108] Trussell J. Optimal diagnosis and medical treatment of male infertility. Seminars in Reproductive Medicine. 2013;**31**:235-236

[109] Jarow JP, Sharlip ID, Belker AM, Lipshultz LI, Sigman M, Thomas AJ. Best practice policies for male infertility. The Journal of Urology. 2002;**167**:2138-2144

[110] Liu D, Baker H. Defective spermzona pellucida interaction: A major cause of failure of fertilization in clinical in-vitro fertilization. Human Reproduction. 2000;**15**:702-708

[111] Guthrie H, Welch G. Effects of reactive oxygen species on sperm function. Theriogenology. 2012;**78**:1700-1708

[112] Singh F, Charles AL, Schlagowski AI, Bouitbir J, Bonifacio A, Piquard F. Reductive stress impairs myoblasts mitochondrial function and triggers mitochondrial hormesis. Biochimica et Biophysica Acta (BBA)-Molecular Cell Research. 2015;**1853**:1574-1585

[113] Mentor S, Fisher D. Aggressive antioxidant reductive stress impairs brain endothelial cell angiogenesis and blood brain barrier function. Current Neurovascular Research. 2017;**14**:71-81

[114] Lamosova D, Jurani M, Greksak M, Nakano M, Vanekova M. Effect of rooibos tea (*Aspalathus linearis*) on chick skeletal muscle cell growth in culture. Comparative Biochemistry and Physiology. 1997;**116**:39-45

[115] Safarinejad MR. Efficacy of coenzyme Q10 on semen parameters, sperm function and reproductive hormones in infertile men. The Journal of Urology. 2009;**182**:237-248

[116] Safarinejad MR, Safarinejad S. Efficacy of selenium and/or N-acetyl-cysteine for improving semen parameters in infertile men: A double-blind, placebo controlled, randomized study. The Journal of Urology. 2009;**181**:741-751

[117] Wong WY, Merkus HM, Thomas CM, Menkveld R, Zielhuis GA, Steegers-Theunissen RP. Effects of folic acid and zinc sulfate on male factor subfertility: A double-blind, randomized, placebo-controlled trial. Fertility and Sterility. 2002;**77**:491-498

[118] Omu AE, Al-Azemi MK, Kehinde EO, Anim JT, Oriowo MA, Mathew TC. Indications of the mechanisms involved in improved sperm parameters by zinc therapy. Medical Principles and Practice. 2008;**17**:108-116

[119] Gupta NP, Kumar R. Lycopene therapy in idiopathic male infertility a preliminary report. International

Urology and Nephrology. 2002; **34**:369-372

[120] Lenzi A, Sgro P, Salacone P, Paoli D, Gilio B, Lombardo F, et al. A placebo-controlled double-blind randomized trial of the use of combined L-carnitine and L-acetylcarnitine treatment in men with asthenozoospermia. Fertility and Sterility. 2004;**81**:1578-1584

[121] Suleiman SA, Ali ME, Zaki ZM, El-Malik EM, Nasr MA. Lipid peroxidation and human sperm motility: Protective role of vitamin E. Journal of Andrology. 1996;**17**:530-537

[122] Hawkes WC, Alkan Z, Wong K. Selenium supplementation does not affect testicular selenium status or semen quality in north American men. Journal of Andrology. 2009;**30**:525-533

[123] Moilanen J, Hovatta O. Excretion of alpha-tocopherol into human seminal plasma after oral administration. Andrologia. 1995;**27**:133-136

[124] Keskes-Ammar L, Feki-Chakroun N, Rebai T, Sahnoun Z, Ghozzi H, Hammami S. Sperm oxidative stress and the effect of an oral vitamin E and selenium supplement on semen quality in infertile men. Archives of Andrology. 2003;**49**:83-94

[125] Kessopoulou E, Powers HJ, Sharma KK, Pearson MJ, Russell ID, Cooke ID. A double-blind randomized placebo crossover controlled trial using the antioxidant vitamin E to treat reactive oxygen species associated male infertility. Fertility and Sterility. 1995;**64**:825-831

[126] Matough FA, Budin SB, Hamid ZA, Louis SR, Alwahaibi N, Mohamed J. Palm vitamin E reduces oxidative stress, and physical and morphological alterations of erythrocyte membranes in streptozotocin-induced diabetic rats. Oxidants and Antioxidants in Medical Science. 2012;**1**:59-68

[127] Therond P, Auger J, Legrand A, Jouannet P. Alpha-tocopherol in human spermatozoa and seminal plasma: Relationships with motility, antioxidant enzymes and leukocytes. Molecular Human Reproduction. 1996;**62**:739-744

[128] Omu AE, Fatinikun T, Mannazhath N, Abraham S. Significance of simultaneous determination of serum and seminal plasma alpha-tocopherol and retinol in infertile men by high-performance liquid chromatography. Andrologia. 1999;**31**:347-354

[129] Lewin A, Lavon H. The effect of coenzyme Q10 on sperm motility and function. Molecular Aspects of Medicine. 1997;**18**:S213-S219

[130] ElSheikh MG, Hosny MB, Elshenoufy A, Elghamrawi H, Fayad S, Abdelrahman S. Combination of vitamin E and clomiphene citrate in treating patients with idiopathic oligoasthenozoospermia: A prospective, randomized trial. Andrology. 2015;**3**:864-867

[131] Moslemi MK, Tavanbakhsh S. Selenium-vitamin E supplementation in infertile men: Effects on semen parameters and pregnancy rate, Int. The Journal of Gene Medicine. 2011;**4**:99-104

[132] Rolf C, Cooper TG, Yeung CH, Nieschlag E. Antioxidant treatment of patients with asthenozoospermia or moderate oligoasthenozoospermia with high-dose vitamin C and vitamin E: A randomized, placebo-controlled, double-blind study. Human Reproduction. 1999;**14**:1028-1033

[133] Jacob RA, Pianalto FS, Agee RE. Cellular ascorbate depletion in healthy men. The Journal of Nutrition. 1992;**122**:1111-1118

[134] Shrilatha B. Muralidhara, early oxidative stress in testis and epididymal sperm in streptozotocin-induced diabetic mice: Its progression and genotoxic consequences. Reproductive Toxicology. 2007;**23**:578-587

[135] Naziroglu M. Enhanced testicular antioxidant capacity in streptozotocin-induced diabetic rats. Protective role of vitamins C and E and selenium. Biological Trace Element Research. 2003;**94**:61-71

[136] Dawson EB, Harris WA, Teter MC, Powell LC. Effect of ascorbic acid supplementation on the sperm quality of smokers. Fertility and Sterility. 1992;**58**:1034-1039

[137] Cyrus A, Kabir A, Goodarzi D, Moghimi M. The effect of adjuvant vitamin C after varicocele surgery on sperm quality and quantity in infertile men: A double blind placebo controlled clinical trial. International Brazilian Journal of Urology. 2015;**41**:230-238

[138] Kumar R, Gautam G, Gupta NP. Drug therapy for idiopathic male infertility: Rationale versus evidence. The Journal of Urology. 2006;**176**:1307-1312

[139] Balercia G, Regoli F, Armeni T, Koverech A, Mantero F, Boscaro M. Placebo-controlled, double-blind, randomized trial on the use of L-carnitine, L-acetylcarnitine, or combined L-carnitine and L-acetylcarnitine in men with idiopathic asthenozoospermia. Fertility and Sterility. 2005;**84**:662-671

[140] Cavallini G, Ferraretti AP, Gianaroli L, Biagiotti G, Vitali G. Cinnoxicam and L-carnitine/acetyl-L-carnitine treatment for idiopathic and varicocele-associated oligoasthenospermia. Journal of Andrology. 2004;**25**:761-772

[141] Sigman M, Glass S, Campagnone J, Pryor JL. Carnitine for the treatment of idiopathic asthenospermia: A randomized, double-blind, placebo-controlled trial. Fertility and Sterility. 2006;**85**:1409-1414

[142] Minamiyama Y, Ichikawa H, Masui T, Kobayashi K, Takemura S, Oka M, et al. Oral administration of reduced coenzyme Q10 ameliorates the endocrine-disrupting chemical-induced sperm toxicity in rats. Free Radical Biology and Medicine. 2016;**100**:S105-S106

[143] Mancini A, Conte B, De Marinis L, Hallgass ME, Pozza D, Oradei A. Coenzyme Q10 levels in human seminal fluid: Diagnostic and clinical implications. Molecular Aspects of Medicine. 1994;**15**:S249-S255

[144] Lafuente R, Gonzalez-Comadran M, Lopez GS, Brassesco R, Carreras R. Coenzyme Q10 and male infertility: A meta-analysis. Journal of Assisted Reproduction and Genetics. 2013;**30**:1147-1156

[145] Roveri A, Casasco A, Maiorino M, Dalan P, Calligaro A, Ursini F. Phospholipid hydroperoxide glutathione peroxidase of rat testis. Gonadotropin dependence and immunocytochemical identification. The Journal of Biological Chemistry. 1992;**267**:6142-6146

[146] Alvarez JG, Storey BT. Lipid peroxidation and the reactions of superoxide and hydrogen peroxide in mouse spermatozoa. Biology of Reproduction. 1984;**30**:833-841

[147] Ursini F, Heim S, Kiess M, Maiorino M, Roveri A, Wissing J. Dual function of the selenoprotein PHGPx during sperm maturation. Science. 1999;**285**:1393-1396

[148] Noack-Fuller G, Beer CD, Seibert H. Cadmium, lead, selenium, and zinc in semen of occupationally unexposed men. Andrologia. 1993;**25**:7-12

[149] Toman R, Hluchy S, Cabaj M, Massanyi P, Roychoudhury S, Tunegova M. Effect of separate and combined exposure of selenium and diazinon on rat sperm motility by computer assisted semen analysis. Journal of Trace Elements in Medicine and Biology. 2016;**38**:144-149

[150] Vezina D, Mauffette F, Roberts KD, Bleau G. Selenium-vitamin E supplementation in infertile men. Effects on semen parameters and micronutrient levels and distribution. Biological Trace Element Research. 1996;**53**:65-83

[151] Powell SR. The antioxidant properties of zinc. The Journal of Nutrition. 2000;**130**:S1447-S1454

[152] Chia SE, Ong CN, Chua LH, Ho LM, Tay SK. Comparison of zinc concentrations in blood and seminal plasma and the various sperm parameters between fertile and infertile men. Journal of Andrology. 2000;**21**:53-57

[153] Harchegani AB, Dahan H, Tahmasbpour E, kaboutarakic HB, Shahriary A. Effects of zinc deficiency on impaired spermatogenesis and male infertility: The role of oxidative stress, inflammation and apoptosis. Human Fertility. 2018:1-12

[154] Hadwa MH, Almashhedy LA, Alsalman AR. Oral zinc supplementation restores superoxide radical scavengers to normal levels in spermatozoa of Iraqi asthenospermic patients. International Journal for Vitamin and Nutrition Research. 2015;**85**:165-173

[155] Kelkel M, Schumacher M, Dicato M, Diederich M. Antioxidant and anti-proliferative properties of lycopene. Free Radical Research. 2011;**45**:925-940

[156] Agarwal A, Sekhon LH. Oxidative stress and antioxidants for idiopathic oligoasthenoteratospermia: Is it justified? Indian Journal of Urology. 2011;**27**:74-85

Chapter 6

Nonenzymatic Exogenous and Endogenous Antioxidants

Ziad Moussa, Zaher M.A. Judeh and Saleh A. Ahmed

Abstract

Nonenzymatic exogenous and endogenous antioxidants play an important role in human health and act as preservatives for cosmetics, pharmaceuticals, and food products. This chapter will discuss the chemical structure and mechanism of action of the most important nonenzymatic small exogenous and endogenous organic molecules that act as antioxidants. The chapter will focus on the structural features, functional groups, properties, biosynthetic origin, and mechanism of action of such antioxidants. It also covers damages that free radicals create and the mechanisms by which they are neutralized by the various antioxidants. The scope of this chapter will be limited to nonenzymatic exogenous and endogenous antioxidants since enzymatic antioxidants have been discussed extensively in several reviews.

Keywords: antioxidants, nonenzymatic, endogenous, exogenous, low-molecular weight antioxidants, mechanism

1. Introduction

Antioxidants are structurally diverse group of small organic molecules and large enzymes that comprise complex systems of overlapping activities working synergistically to enhance cellular defense and to combat oxidative stress resulting from various reactive oxygen species (ROS) and reactive nitrogen species (RNS) [1]. The former substances are byproducts of metabolism and are ironically produced from oxygen, an indispensable element for life. Many of these reactive species are free radicals possessing one or more unpaired electrons and as such rendered highly reactive. The reactive species generated in cells include hydrogen peroxide (H_2O_2), hypochlorous acid (HClO), the hydroxyl radical (·OH), the superoxide anion radical (O_2^-), the nitric oxide radical (NO·), and the lipid peroxyl radical (LOO·) [2, 3]. The term antioxidants may refer to either industrial chemicals that may be added to products to combat oxidation or to natural products that are found in foods and tissue. While the former act as preservatives for cosmetics, pharmaceuticals, and food products, the latter play an important role in human health as well. There are many reactive oxygen species conducting unwanted oxidation reactions in a variety of cell and tissue sites [4]. Likewise, each antioxidant targets specific types of ROS and provides protection in distinct environments. Antioxidants reduce reactive oxygen species which otherwise participate in oxidation reactions that can generate free radicals and cause damage to cellular components such as DNA, proteins, carbohydrates, and lipids [4]. It is noted, however, that reactive oxygen species mediate certain cellular functions like redox signaling and gene expression as well as defend against pathogens [5, 6]. Thus, the role of antioxidant systems is not to eliminate oxidants completely, but instead maintain them at an optimum level. Despite the presence of the antioxidant defense mechanism to counteract oxidative stress, damage due to oxidation has a cumulative effect and has been implicated in several chronic conditions and disease states such as cancer [7], cardiovascular disease [8], and neurodegenerative disorders [9]. Antioxidant compounds and antioxidant enzyme systems display synergistic and interdependent effects on one another. Antioxidants found in nature can be classified in a number of ways. Based on their activity, they can be classified as enzymatic and nonenzymatic antioxidants (phytochemicals and vitamins). While antioxidant enzymes like superoxide dismutase (SOD) [10], glutathione peroxidase (GPx) [11], glutathione reductase (GSR) [12], peroxiredoxin I-IV and catalases (CAT) [13] are macromolecules, the vast majority of the remaining natural antioxidants classified as phytochemicals and vitamins are relatively smaller organic molecules with low molecular weights [14, 15]. Antioxidants have also been categorized as water-soluble or fat-soluble molecules.

This chapter will highlight the chemical structures and mechanism of action of important nonenzymatic small exogenous (natural) and endogenous (synthetic/physiological) organic molecules that act as antioxidants in plants and animals. The antioxidants described in this chapter are among the most important, although certainly they are not the only ones known. Special focus on the structural features, functional groups, properties, biosynthetic origin, and mechanism of action will be undertaken with special coverage of damages that free radicals create and the mechanisms by which they are neutralized by the various antioxidant molecules.

2. Enzymatic versus nonenzymatic antioxidants

Based on their activity, antioxidants are classified as enzymatic and nonenzymatic antioxidants. While enzymatic antioxidants [10–13] function by converting oxidized metabolic products in a multi-step process to hydrogen peroxide (H_2O_2) and then to water using cofactors such as iron, zinc, copper, and manganese, nonenzymatic antioxidants intercept and terminate free radical chain reactions. Examples of natural nonenzymatic antioxidants are vitamin E, A, C, flavonoids, carotenoids, glutathione, plant polyphenols, uric acid, theaflavin, allyl sulfides, curcumin, melatonin, bilirubin, and polyamines [14, 15]. Some of these antioxidants are water-soluble and predominantly found in the cytosol or cytoplasmic matrix, while others are liposoluble and are present in cell membranes. The enzymatic antioxidants and their mechanism of action have been discussed extensively in several review articles [16–18]. The scope of this chapter will be limited to nonenzymatic exogenous and endogenous antioxidants.

3. Generation of free radicals in living organisms

The production of ROS in biological systems occurs during oxygen metabolism and plays an important role in homeostasis and cell signaling [5]. However, under conditions of environmental stress, the concentration of ROS can increase significantly and inflict damage on cell structures. The generation of ROS begins with the reduction of molecular oxygen with NADPH to produce the superoxide anion radical (O_2.–), a precursor to most remaining reactive oxygen and nitrogen species (**Figure 1**). Subsequent dismutation of two molecules of the superoxide anion catalyzed by the enzyme superoxide dismutase (SOD) generates oxygen and hydrogen peroxide. The latter in turn may undergo partial reduction to hydroxyl radical through the Fenton reaction or alternatively *via* the Haber-Weiss process [19]. While hydrogen peroxide is more damaging to DNA, the hydroxyl radical is highly reactive and turns biomolecules into free radicals, thus perpetuating a free radical chain reaction. Hydrogen peroxide may also be converted to the potent oxidant hypochlorous acid in the presence of the chloride ion, an omnipresent species. This transformation is catalyzed by the enzyme myeloperoxidase (MPO). Reaction of HOCl with H_2O_2 regenerates chloride ion and produces singlet oxygen as yet another ROS. On the other hand, RNS such as nitric oxide (NO•) are produced by the enzyme nitric oxide synthase (NOS) starting from the precursor L-arginine [20]. Nitric oxide functions as a superoxide quencher forming peroxynitrite (ONOO−), a strong oxidant that reacts indiscriminately with biological targets. Further, it may disintegrate into a pair of hydroxyl and nitric dioxide radicals and cause damage through such species (**Figure 1**).

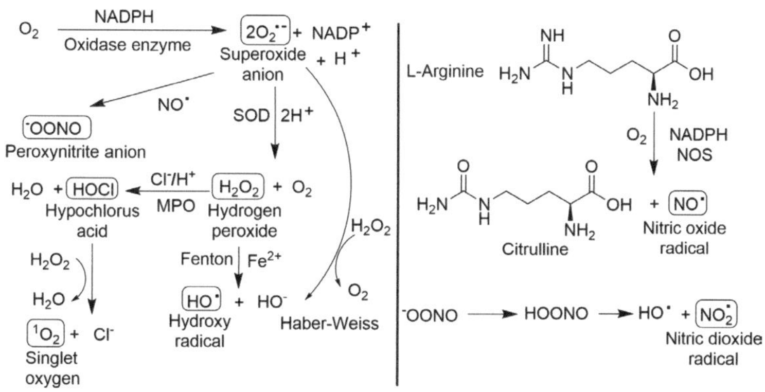

Figure 1.
Generation of ROS and RNS in living species.

4. Damaging chemical reactions of free radicals in living organisms

4.1 Free radical damage to the deoxyribose moiety of DNA

The highly reactive hydroxyl radical (•OH) reacts with the sugar moiety of DNA causing structural modification and strand breaks by a variety of mechanisms [21]. The OH radical reacts with the 2′-deoxyribose sugar residue in DNA by abstracting $H^•$ from all its carbon atoms forming five carbon-centered radicals. The H4′ and H5′ atoms are more accessible to $H^•$ abstraction by the OH radical than the H1′, H2′, and H3′. The C4′ C-centered radical appears to be the major radical generated by $H^•$ abstraction from 2′-deoxyribose in DNA [22]. These radicals undergo further reactions, producing a variety of 2′-deoxyribose oxidative adducts. While some products detach from DNA, others remain tethered as end groups of fragmented DNA strands [22]. In the absence of oxygen and as depicted in **Figure 2**, one of the byproducts formed from C4′-radical of 2′-deoxyribose as an end group of a severed DNA strand is 2,5-dideoxypentose-4-ulose. The product is formed by heterolytic cleavage of the phosphate group at C5′ to give a C4′/C5′-radical cation which in turn undergoes hydration and subsequent one-electron reduction and base elimination (**Figure 2**). Other products formed from the C4′ radical include 2-deoxypentose-4-ulose and 2,3-dideoxypentose-4-ulose. However, in the presence of oxygen, rapid addition of O_2 to the C4′-radical forms a peroxyl radical which undergoes a series of fragmentation reactions yielding 3′-phosphoglycolate as an end group [23].

Oxidation of the C1′, C2′, and C5′ radicals yields products such as 2-deoxypentonic acid lactone, erythrose, 2-deoxytetradialdose, and 5′ aldehyde [22].

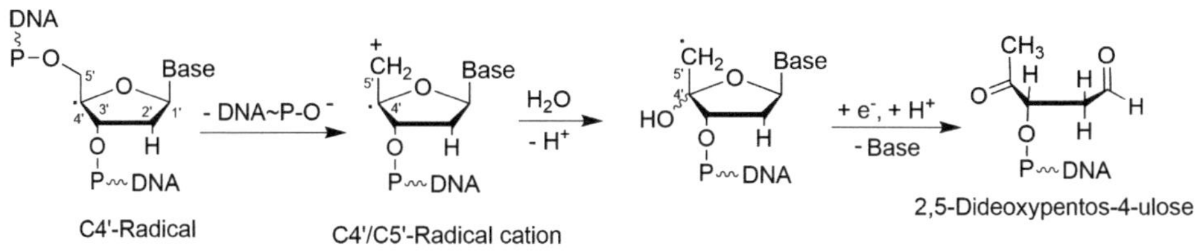

Figure 2.
Mechanism of product formation from reactions of the C4′-radical of 2′-deoxyribose, leading to 2,5-dideoxypentose-4-ulose as an end group of a broken DNA strand.

4.2 Free radical damage to DNA bases

Besides reacting with the sugar moiety of DNA, the highly reactive hydroxyl radical (·OH) reacts with the heterocyclic bases guanine (**Figure 3**), thymine, and cytosine, causing free radical-induced DNA damage by several different pathways. Guanine, however, possesses the lowest reduction potential (1.29 V) among the four DNA bases, rendering the motif the best electron donor and prone to preferential oxidization [24]. The hydroxyl radical reacts with the C4-, C5-, and C8-positions of guanine and to a lesser extent with the C2-position, generating a plethora of products. Interestingly, the HO-adduct radicals generated from the addition reactions of HO· may exhibit reducing or oxidizing properties (redox ambivalence), yielding the relevant products accordingly. Hence, while the C5-OH– and the C8-OH–adduct radicals are reducing, the C4-OH–adduct radical is predominantly oxidizing. The last two adduct radicals form in yields of 17% and 65–70%, respectively, whereas the yield of the C5-OH– adduct radical is lower than 10% [25]. Although formed in relatively lower yields, the C8-OH–adduct radical produces the major byproducts of guanine reactions (**Figure 3**). Thus, as shown in **Figure 3** and following reaction of the hydroxyl radical with the C-8 position of guanine, one-electron oxidation of the resulting C8-OH–adduct radical yields the enol form of 8-hydroxyguanine which undergoes tautomerization to generate the predominant keto form [21]. The latter may also form via a pathway involving 1,2-hydride-shift and subsequent oxidation of the C8-OH–adduct radical. The 1,2-hydride-shift radical product may also undergo single electron reduction, followed by ring opening reaction to form 2,6-diamino-4-hydroxy-5- formamidopyrimidine. The preceding radical damage to DNA has been directly correlated to several disease states such as genetic mutation, atherosclerosis, Alzheimer's disease, and the aging process [26, 27].

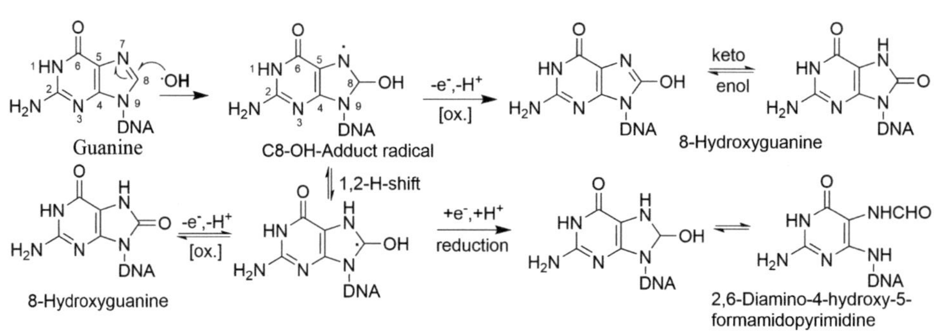

Figure 3.
Oxidative and reductive product formation from reactions of the C8-OH–adduct radical of guanine.

4.3 Free radical damage to polyunsaturated fatty acid groups of cell membranes

While free radicals react with all major classes of biomolecules, peroxidation of the polyunsaturated fatty acid groups (PUFA) of cell membranes comprises the main target of oxidative damage, resulting in a destructive self-propagating chain reaction. The general mechanism of PUFA peroxidation involves abstraction of hydrogen from a lipid molecule (LH) by an initiator (R·) to generate a carbon-based free radical (L·) which reacts rapidly with molecular oxygen to form the peroxyl radical (LOO·) known to propagate the chain reaction (**Figure 4**). As such, the peroxyl radical reacts with PUFA moieties, producing lipid hydroperoxides (LOOH) and perpetuating the chain reaction. The hydroperoxides can further dissociate to dangerous radical species like bioactive aldehydes which inflict damage on other cellular components. Lipid hydroperoxidation has been linked to a number of physiological conditions and tissue injuries [28].

$$LH + R^{\bullet} \longrightarrow RH + L^{\bullet} \xrightarrow{O_2} LOO^{\bullet} \xrightarrow{LH} LOOH + L^{\bullet}$$

Figure 4.
General process of lipid peroxidation.

5. Regulation of free radicals with nonenzymatic small natural exogenous antioxidants

5.1 Vitamins

5.1.1 Vitamin E

Vitamin E is a collection of optically active methylated phenolic compounds comprising four tocopherols and four tocotrienols [29] where α-tocopherol is the most common and biologically active species (**Figure 5**) [30]. The structures feature two primary parts: a densely substituted polar chromanol aromatic ring and a lipophilic long polyprenyl side chain. The main chemical structural difference between different forms of Vitamin E is that tocotrienols feature unsaturated isoprenoid hydrocarbon side chains with three carbon-carbon double bonds versus saturated isoprenoid side chains for tocopherols. Within each group, the vitamers are differentiated by the number and positions of the methyls in the chromate ring. The polyprenyl precursor for the biosynthesis of tocopherols and tocotrienols is phytyl pyrophosphate (PPP) and geranylgeranyl pyrophosphate (GGPP), respectively [31]. Vitamin E is biosynthesized though the shikimate pathway, and while α-tocopherol and α-tocotrienol are considered structurally unique, the remaining compounds in each class are constitutional isomers. The presence of three stereogenic centers (position C_2 of the chromate ring, position C_4 and C_8 of the phytyl side chain) produces 8 different stereoisomers (four pairs of enantiomers) depending on the position and orientation of the groups in each of the chiral centers. Since the discovery of vitamin E in 1920, it has been shown to be the most powerful membrane-bound antioxidant utilized by cells to scavenge reactive nitrogen and oxygen species with consequent disruption of oxidative damage to cell membrane phospholipids during cellular lipid peroxidation of the polyunsaturated fatty acids (PFA) and low-density lipoprotein (LDL) [32]. The antioxidant is liposoluble and localized to cell membranes. Vitamin E functions by reducing lipid peroxyl radicals ($LOO^{\cdot}$) by transferring the phenolic hydrogen atom of the chroman ring (**Figure 5**), resulting in a relatively stable and unreactive resonance-stabilized tocopheroxyl radical which is unable to trigger further lipid peroxidation itself. The α-tocopherol radical can be reduced back to the original active α-tocopherol form by ascorbic acid or coenzyme Q10 [33, 34]. Alternatively, it may quench a second peroxyl radical where the resulting tocopheryl peroxide eliminates a peroxide leaving group, forms a hemiketal after reacting with water, and lastly hydrolyses to the tocopherolquinone. This is an essential foundation and benchmark of a good antioxidant. The synergistic antioxidation interactions between vitamin E and the ascorbate ion of vitamin C position the former at the forefront of the anti-radical defense system. Vitamin E is exogenous and hence is essential and must be obtained through diet in small amounts since the organism cannot synthesize it. Its biosynthesis is restricted to plants, photosynthetic algae, and certain cyanobacteria. Although vitamin A deficiency is rare, the most frequent manifestations of its lack comprise a number of disorders and disease states which include encephalomalacia, exudative diathesis, muscular dystrophy, and ceroid pigmentation. α-Tocopherol exhibits the highest bioactivity (100%), with the relative activities of β-, γ-, and δ-tocopherols being 50, 10, and 3%, respectively [35].

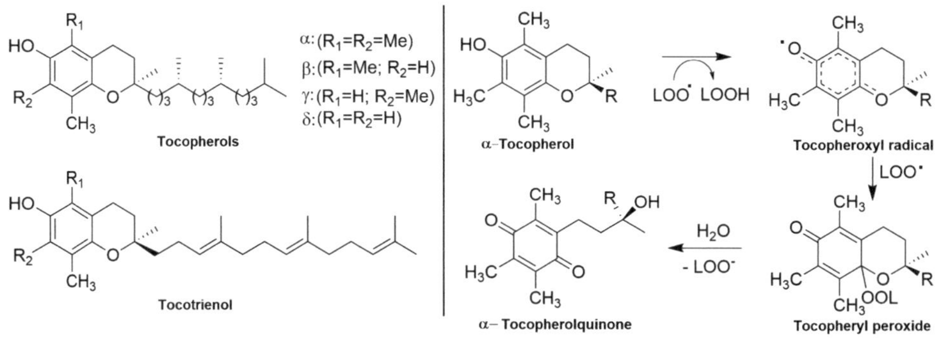

Figure 5.
Chemical structures of the tocopherols and tocotrienols that comprise vitamin E and termination of lipid peroxidation with α-tocopherol.

5.1.2 Vitamin A

Vitamin A, just like vitamin E, is a term that designates a family of unsaturated liposoluble organic compounds that include retinol, retinal, retinoic acid, and retinyl palmitate, and many provitamin A carotenoids such as beta-carotene (**Figure 6**). All forms share a beta-ionone ring to which an isoprenoid tether known as retinyl group is attached. It is noteworthy that both features are essential for vitamin A activity. The common chemical structure is a diterpene ($C_{20}H_{32}$) where the various molecular forms differ by the terminal side chain functional group. Thus, retinol contains a hydroxyl group, retinal contains an aldehyde function, retinoic acid has a terminal carboxylic acid group, and retinyl palmitate bears an ester moiety. The discovery of the antioxidant activity of vitamin A dates back to 1932 when Schmitt and Monaghan reported that vitamin A prevents lipid rancidity [36]. Several reviews outlining the antioxidant role and metabolic functions of vitamin A have appeared in the literature [37, 38]. Besides eliminating free radicals, it plays a major role in maintaining good vision. The aldehyde form of vitamin E is required by the retina to form the light-absorbing molecule rhodopsin necessary for both color and scotopic vision [39]. On the other hand, the fully irreversibly oxidized form of retinol functions in a very different way as a growth factor for epithelial and other types of cells [38]. As an antioxidant, vitamin A scavenges lipid peroxyl radicals (LOO•) according to the mechanism shown in **Figure 6**. Thus, by trapping the peroxyl radical through an addition reaction to the beta-ionone ring of retinol, the resultant tertiary and highly conjugated trans-retinol carbon radical intermediate is relatively stable and under normal conditions is not reactive enough to induce further lipid peroxidation itself. However, the intermediate may continue reacting with lipid peroxyl radicals or molecular oxygen to produce a bis-peroxyl adduct or retinol-derived peroxyl radical, respectively. Alternatively, it may eliminate LO radical and oxidizes to 5,6-retinol epoxide [15].

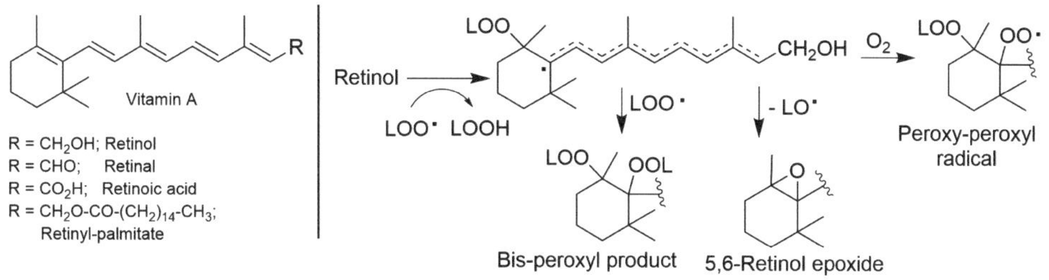

Figure 6.
Chemical structure of vitamin A and termination of lipid peroxidation with retinol.

5.1.3 Vitamin C

Vitamin C (L-ascorbic acid) is an optically-active hydrosoluble free radical scavenger that bears a highly acidic hydroxyl group (pKa = 4.2) known to be completely ionized at neutral pH [35, 40]. Thus, the acidic vitamin readily loses a proton from the 3-hydroxyl group affording a resonance-stabilized ascorbate anion ($AscH^-$) (**Figure 7**). The unusual acidity of the alcohol is related to the presence of two conjugated double bonds which stabilize the deprotonated monoanionic conjugate base. Furthermore, these same electronic factors impart stability to the radical form of vitamin C when it undergoes one electron oxidation by lipid radicals to generate the ascorbate radical (**Figure 7**), a much less reactive species than most other free radicals. As such, vitamin C is able to assume the role of a free-radical scavenger. The low standard 1-electron reduction potential (282 mV) renders vitamin C an excellent electron donor. As well, at low ascorbate concentrations, it may function as a pro-oxidant reducing agent and is able to reduce redox-active copper and iron metals. Vitamin C is therefore required as a cofactor for a number of metabolic processes that mediate essential biological functions in all animals and plants [41]. The structure features a chiral 3,4-dihydroxyfuran-2(5*H*)-one ring and a 1,2-dihydroxyethyl tether containing another stereogenic center. The 6-carbon ketolactone is structurally related to glucose. Although four stereoisomers are expected depending on the position of the substituents around the stereogenic centers, only the L-enantiomer exhibits antioxidant capacity in biological systems, both in vitro and in vivo. While vitamin C is biosynthesized by nearly all animals, humans comprise a notable exception. Consequently, it is an essential nutrient and must be obtained through dietary means. In biological species, the vitamin exists in the protonated form at low pH, but in media with pH above 5, it is found in the dissociated ascorbate form [42]. This species is a 2-electron donor and gets oxidized to a molecule of dehydroascorbate (DHA) which does not have any antioxidant capacity. However, regeneration of the ascorbate from DHA is possible by the addition of two electrons and has been proposed to be carried out by oxidoreductase [43]. In animals, the biosynthesis of ascorbic acid is carried out by several enzymes in the liver from glucose [42], by a synthetic route which initially involves oxidation to D-glucuronic acid via uridine diphosphate (UDP) derivatives. Subsequent reduction of the open-chain aldehyde form of D-glucuronic acid to the primary alcohol (L-gulonic acid), lactone formation between the carboxyl and 4-hydroxyl group, oxidation of the secondary hydroxyl function to a carbonyl, and subsequent enolization result in L-ascorbic acid. The latter, specifically in the ascorbate form, acts as a reducing agent, donating electrons to lipid radicals in order to terminate the lipid peroxidation chain reaction (**Figure 7**). Another main function of Vitamin C as an antioxidant is to regenerate vitamin E (HO-tocopherol) from its oxidized form (.O-tocopherol) back to its active state by reducing vitamin E radicals formed when vitamin E scavenges oxygen radicals. The recycling of vitamin E is carried out in cell membranes in conjunction with glutathione (GSH) or other sacrificial reductants [33, 34]. Likewise, vitamin C acts as an antioxidant and reducing agent by donating electrons to various enzymatic and nonenzymatic reactions. It reduces the transition metal ions of several biosynthetic enzymes, thus preventing biological oxidation of macromolecules. In plants, vitamin C is a substrate for the enzyme ascorbate peroxidase which catalyzes the reduction of toxic hydrogen peroxide (H_2O_2) to water (H_2O) [44]. Currently, this vitamin is the most widely employed vitamin in drugs, premedication, and dietary supplements worldwide.

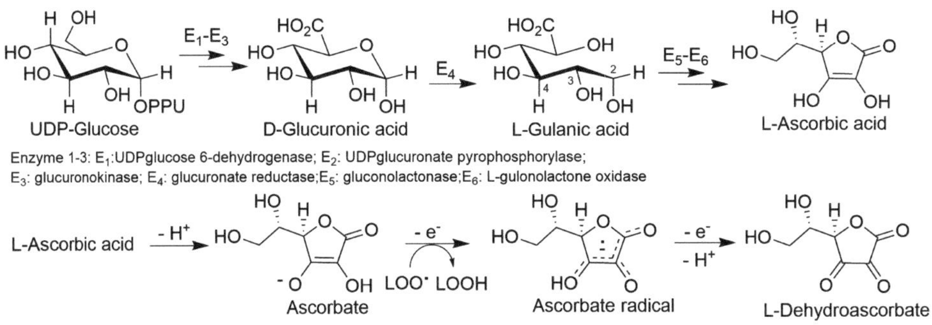

Figure 7.
Biosynthesis, chemical structure, and reduction mechanism of ascorbic acid.

5.2 Flavonoids

Flavonoids are exogenous antioxidants displaying rich structural diversity and are ubiquitous in plants and certain photosynthetic organisms. More than 8000 of these benzo-γ-pyran derivatives have been identified and characterized [45, 46]. The general structure features a C6-C3-C6 15-carbon flavone skeleton, which comprises two phenyl rings (A and B) linked by a heterocyclic ring (C) (**Figure 8**). Flavonoids have been classified into flavones, flavanones, flavanols, flavonols, and anthocyanins. While flavones have a double bond between C2 and C3, flavanones have a saturated C2–C3 bond. Compared to flavones, the corresponding flavonols have an additional hydroxyl group at the C3 position while flavonols are C2-C3 saturated analogs of flavonols. Flavonoid groups are differentiated based on the number of hydroxyl and other substituents on the phenyl rings [47].

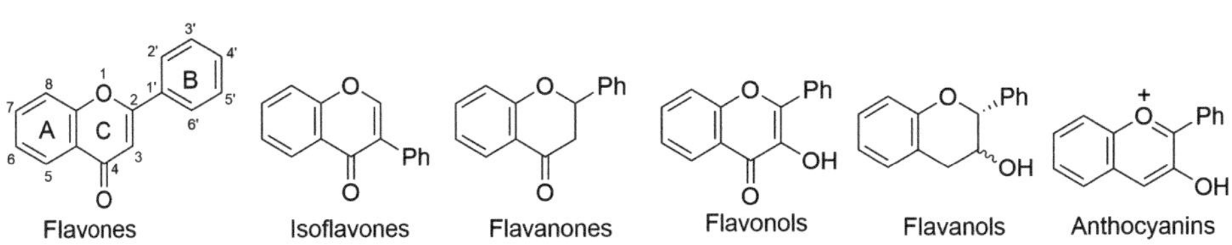

Figure 8.
General skeletal structure of the various flavonoid classes.

Quercetin (3,5,7,3′,4′–pentahydroxyflavone) (**Figure 9**) is the most ubiquitous polyphenolic flavonoid known to prevent oxidative damage to DNA oligonucleotides brought about by H_2O_2, HO·, and $O_2^{\cdot-}$. On the other hand, anthocyanidin is a strong inhibitor of lipid oxidations. Thus, as shown in **Figure 9**, the antioxidant mechanism of lipid peroxyl radicals scavenging capability of anthocyanidin is based on its hydrogen radical donation ability from the *p*-hydroxyl group of ring B to generate a resonance-stabilized anthocyanidin radical incapable of participating in other radical reactions. In addition, the effectiveness of anthocyanidin in inhibiting lipid peroxidation has been correlated to their metal-ion chelating power [48, 49]. In particular, the ortho-dihydroxy groups in the B-ring confer upon this class of compounds antiperoxidative properties [50]. However, phenolic compounds can also act as prooxidants if present in high concentrations with metal ions and high pH [47].

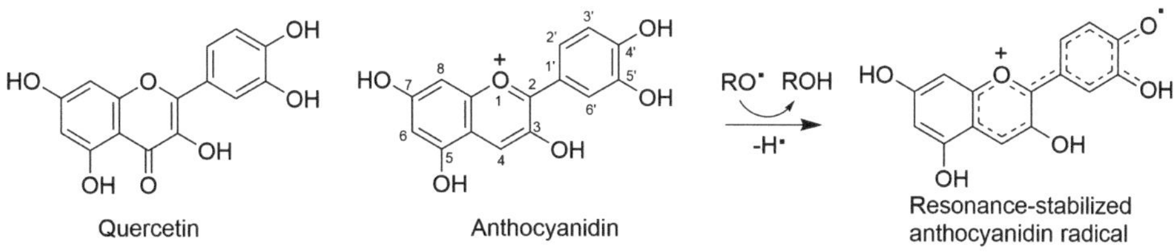

Figure 9.
Structure of quercetin and mechanism of radical scavenging activity of anthocyanidin.

5.3 Carotenoids

Carotenoids, also known as tetraterpenoids, are a group of phytonutrients produced by plants and algae, as well as some bacteria and fungi [51]. The long unsaturated hydrocarbon alkyl chain renders carotenoids highly liposoluble. Hence, they play a key role in the protection of lipoproteins and cellular membranes from lipid peroxidation and exhibit particularly efficient scavenging capacity against peroxyl radicals as compared to any other ROS and they are known to be the most common lipid-soluble antioxidants [52, 53]. Over 1100 carotenoids have been identified and classified primarily into two groups: the oxygen-containing xanthophylls and those that are purely hydrocarbons, carotenes (**Figure 10**). Biosynthetically, all carotenoids are tetraterpenes comprising 40 carbon atoms which are produced from eight isoprene units. The structural backbone consists of isoprenoid units biosynthesized either by head-to-tail or by tail to-tail process. The basic building blocks of carotenoids are isopentyl diphosphate (IPP) and dimethylallyl diphosphate (DMAPP) which produce the major carotenoid precursor geranylgeranyl pyrophosphate (GGPP) [54]. GGPP undergoes several different reactions within the carotenoid biosynthetic pathway to afford carotenes or xanthophylls. Carotenoids reduce the peroxyl radicals to form a resonance-stabilized carbon-centered radical product. Lycopene and carotene are the most prominent and potent carotenoid antioxidants. The former is notably a strong singlet oxygen quencher due to the high number of conjugated *trans*-configuration double bonds present in the structure. In general, the extended conjugated system in carotenoids is strongly-reducing, facilitating abstraction of hydrogen atoms from the allylic positions to this conjugation, as well allowing free-radical addition reactions to proceed with ease. Lycopene for instance reduces peroxyl radicals through electron transfer to afford an unreactive resonance stabilized carbon-centered radical (**Figure 10**).

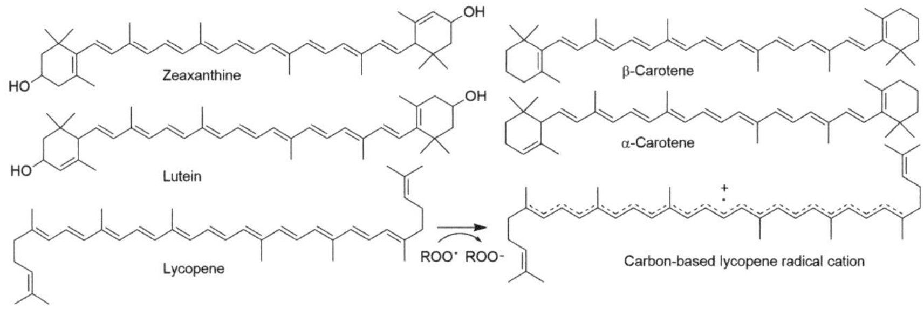

Figure 10.
Structure and radical scavenging mechanism of some prominent xanthophyll and carotene carotenoids.

5.4 Hydroxycinnamic acids

Hydroxycinnamic acids (hydroxycinnamates) are a class of phenylpropanoids possessing a C_6–C_3 skeleton. These compounds are hydroxy derivatives of cinnamic acid which is their common biosynthetic precursor. Mechanistically, bimolecular elimination of ammonia from the side chain of L-Phenylalanine generates *trans*-(*E*)-cinnamic acid (**Figure 11**). A subsequent cytochrome P-450-dependent direct hydroxylation reaction of cinnamic acid mediated by cinnamate 4-hydroxylase enzyme (E2) produces the first member of this class, *p*-coumaric acid. The substitution patterns of the remaining cinnamic acids are constructed sequentially by further hydroxylation and methylation reactions, which is typical of shikimate pathway metabolites. Hence, direct hydroxylation of *p*-coumaric acid mediated by *p*-coumarate 3-hydroxylase enzyme (E3) generates caffeic acid. Subsequent methylation of the latter by caffeic acid *O*-methyltransferase (E4) produces ferulic acid. Hydroxylation of ferulic acid by ferulate-5-hydroxylase (E5), a cytochrome P450-dependent monooxygenase enzyme, followed by methylation with SAM produces the last member, sinapic acid. As chain-breaking antioxidants, hydroxycinnamic acids prevent oxidation of LDL, although in varying efficiencies, depending on their standard one-electron reduction potential, hydrogen or electron donating ability, and the capacity to delocalize and stabilize the resulting phenoxyl radical within their structural framework [55, 56]. The antioxidant activity of the derivatives is correlated with the methylation and hydoxylation substitution pattern of the benzene ring. Thus, the antioxidant efficiency of the hydroxycinnamate conjugates on human LDL oxidation has been found to increase in the order of *p*-coumaric acid, ferulic acid, sinapic acid, and caffeic acid [57]. The general mechanism of free radical scavenging by which these antioxidants act involves donation of a *p*-hydroxyl hydrogen atom to ROS and generation of resonance stabilized carbon-based radical. Additionally, the presence of *ortho*-dihydroxyl groups allows metal-ion chelation much like flavonoids and enhances their antioxidant capacity against lipid peroxidation.

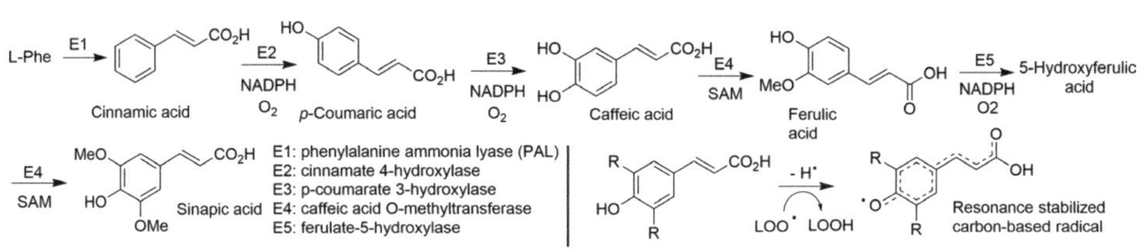

Figure 11.
Structures, biosynthesis, and free radical scavenging mechanism of hydroxycinnamic acids.

5.5 Other natural exogenous antioxidants (allyl sulfides & curcumin)

Allicin (diallyl thiosulfinate), a compound mainly found in garlic, and curcumin are biologically active compounds possessing antioxidative properties. The active form responsible for the antioxidant activity of allicin is 2-propene-sulfenic acid [58], formed via a cope elimination reaction of the former precursor (**Figure 12**) [59, 60]. The radical-scavenging mechanism of allicin involves H-atom abstraction by a peroxyl radical from the sulfenic acid residue [61, 62]. The bis-α,β-unsaturated β-diketone, curcumin, is a liposoluble free radical scavenger that displays remarkable chain breaking ability similar to that of vitamin E [63]. As shown in **Figure 11**, the methylene group of the β-diketone residue and the phenolic hydroxyl (OH) function are sites that can transfer electrons or H-atoms to quench free radicals and generate extended resonance-stabilized carbon- or oxygen-centered radicals. The phenoxyl radical, which has been credited for the antioxidative properties of curcumin [64], generates a quinone methide as it moves through the carbon framework and reacts with molecular oxygen to produce a peroxyl radical. Subsequent reduction of the peroxyl radical and dehydration of the resulting hydroperoxide, followed by rearrangement into a spiro-epoxide and hydrolysis, give the final bis-cyclopentadione product.

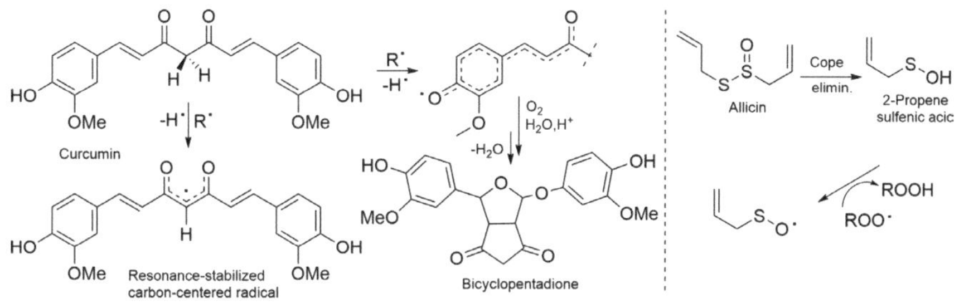

Figure 12.
Structures and radical-scavenging activities of curcumin and allicin.

6. Regulation of free radicals with nonenzymatic small endogenous (synthetic/physiological) antioxidants

6.1 Uric acid

Uric acid (UA) is a hydrophilic antioxidant generated during the metabolism of purine nucleotides and accounts nearly for 66% of the total oxygen scavenging activity in the blood serum. Mammals and humans are capable of producing UA, making it the most predominant aqueous antioxidant present in humans [65, 66] with an approximate blood level of 3.5–7.5 mg/dL. UA is a strong electron donor and a selective scavenger of peroxynitrite ($ONOO^-$), requiring the participation of ascorbic acid and thiols in its cycle for complete scavenging of such species [67, 68]. Peroxynitrite is formed by the reaction between nitric oxide (·NO) and superoxide radical ($O_2^{\cdot-}$) (**Figure 1**) and has been implicated in many pathologies. Besides scavenging peroxynitrite, UA reacts with hydroxyl radicals, singlet oxygen, lipid peroxides, and hypochlorous acid, itself getting converted to innocuous chemical species like urea and allantoin. Furthermore, it has been implicated in scavenging carbonate ions ($CO_3^{\cdot-}$) and nitrogen dioxide ($NO_2^{\cdot}$) [69], and in complexation with copper and iron ions, resulting in the inhibition of deleterious free radical reactions like the Fenton and the Haber-Weiss reactions [65]. Some have suggested that UA does not directly scavenge peroxynitrite since UA cannot compete for the reaction of peroxynitrite with CO_2. The antioxidant effect of uric acid may thus be related to the scavenging of the radicals CO_3·– and NO_2· which are formed from the reaction of peroxynitrite with CO_2 [67]. As shown in **Figure 13**, UA displays a keto-enol tautomerism where the enol form predominantly exists as the monobasic urate anion at physiological pH [70]. The complete scavenging of peroxynitrite requires the presence of ascorbic acid and thiols whereby the urate anion is regenerated following reduction of the urate free radical with ascorbate ($AscH^-$). ESR studies on UA radical production by hydrogen atom abstraction provided evidence that the unpaired electron resides primarily on the five-membered ring of the purine structure. The radical was described as a delocalized π radical as the odd electron showed spin density on all four nitrogen atoms [71].

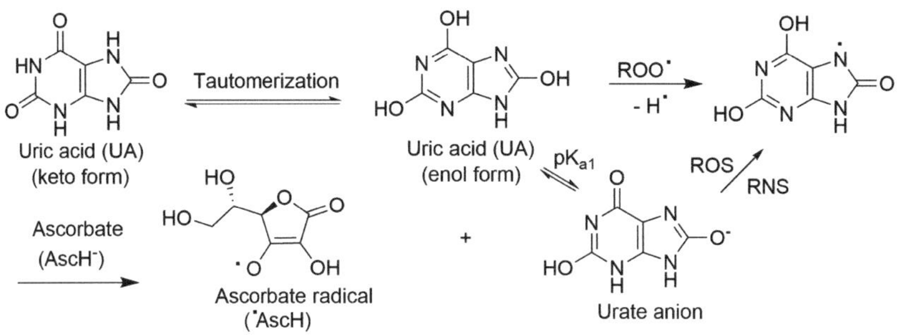

Figure 13.
Chemical structure and radical scavenging mechanism of uric acid.

6.2 Glutathione

Glutathione (GSH) is present in all plant and animal cells and comprises three amino acids: glycine, cysteine, and glutamic acid. It is mainly synthesized in the liver [72] and exists in several redox forms, among which the most predominant is the reduced glutathione. GSH is a hydrosoluble antioxidant present in high cellular concentrations (1–10 mM) in the nucleus, mitochondria, and cytoplasm. GSH is involved in several lines of defense against ROS. First, the thiol group confers GSH with the ability to protect other thiol functions in proteins against oxidative damage [73]. Thiol groups (-SH) are widespread and highly reactive chemical entities in cells. They complex with metal ions, participate in oxidation reactions by getting oxidized themselves to sulfonic acids, and form thiol radicals and disulfides [74]. As an antioxidant, GSH reduces ROS during the enzymatic and nonenzymatic reactions. It regenerates other oxidized antioxidants like vitamin C and vitamin E [75] and is involved in the repair of lipids damaged in peroxidation processes and in the maintenance of sulfhydryl moieties of proteins in the reduced form [76, 77]. GSH functions in conjunction with three groups of enzymes to maintain an intracellular reducing environment and combat excessive formation of harmful ROS. These enzymes are glutathione peroxidase (GSHPx), glutathione reductase (GR), and glutathione oxidase (GOx). Glutathione peroxidase (GSHPx) is a selenium-containing enzyme that mediates catalytic reduction of peroxides using GSH as a sacrificial reductant [78]. The enzyme is a tetramer featuring a selenocysteine residue in each subunit [11]. The oxidation-reduction chemistry of the selenol functional group found in each selenocysteine is responsible for the activity of GSHPx, and the catalytic cycle is displayed in **Figure 14** [79]. In the first step, the selenol functional group (EnzSeH) gets oxidized by the peroxide to the corresponding selenenic acid (EnzSeOH). The thiophilic acid reacts with GSH to generate a selenenyl sulfide intermediate (EnzSeSG) which is highly reactive and is susceptible to nucleophilic displacement at the sulfur atom. Thus, attack by a second molecule of GSH at the sulfur atom regenerates the original selenol and eliminates oxidized glutathione (GSSG) as a byproduct. The latter is recycled back to GSH in an NADPH-dependent reduction process mediated by glutathione reductase (GR). GSH is also a substrate for glutathione oxidase (GOx) which catalyzes the reduction of oxygen to hydrogen peroxide and GSSG.

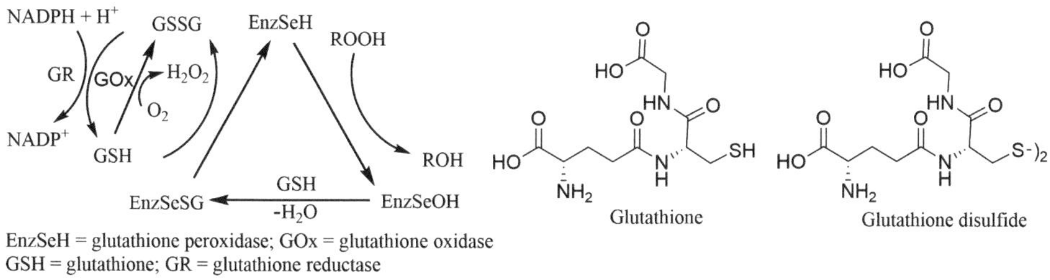

Figure 14.
Structure and role of glutathione (GSH) in the catalytic cycle of glutathione peroxidase (GSHPx), glutathione reductase (GR), and glutathione oxidase (GOx).

6.3 Melatonin

Since its discovery in 1993, melatonin's ability to reduce oxidative stress induced in all cells and organs by both oxygen- and nitrogen-based radicals has been reported in over one thousand publications. The structure of this endogenous antioxidant features an indoleamine and is biosynthesized in animals from L-tryptophan, an intermediate product of the shikimate pathway [80]. The biosynthetic process includes hydroxylation, decarboxylation, acetylation, and a methylation (**Figure 15**). Melatonin, which is produced mainly by the pineal gland in the brain [81], indirectly reduces free radical formation primarily through a process known as radical avoidance by stimulating the expression of endogenous antioxidant enzymes that metabolize reactive species and maintain redox homeostasis within cells [82]. These include superoxide dismutase (SOD), glutathione peroxidase (GSHPx), glutathione reductase, and catalase. In addition, it induces the synthesis of the antioxidant glutathione and inhibits certain enzymes that normally produce free radicals like nitric oxide synthase (generates $NO^{\bullet}$). Melatonin can also directly scavenge free radicals along with several of its metabolites that are formed during radical neutralization [83, 84]. For example, it is a very effective scavenger of the hydroxyl radical, singlet oxygen, peroxynitrite anion, and nitric oxide. Interestingly, melatonin has been shown to exhibit double the activity of vitamin E and ranks among as the most effective lipophilic antioxidant.

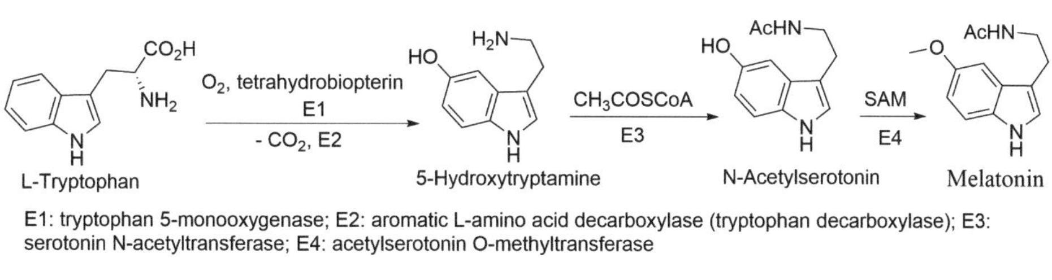

Figure 15.
Structure and biosynthesis of melatonin.

6.4 Bilirubin

Bilirubin (BIL) is an endogenous antioxidant produced from the enzymatic degradation of hemoglobin and other heme proteins (**Figure 16**). The process involves oxidative cleavage, catalyzed by the enzyme heme oxygenase, of one porphyrin exocyclic double bond of a heme residue of hemoglobin to generate biliverdin. Subsequent enzymatic reduction of biliverdin by biliverdin reductase yields bilirubin. This process is reversible and the oxidation of bilirubin by lipophilic ROS results in the formation of biliverdin. Notable structural features of bilirubin include an open chain of four connected pyrrole rings and a *Z,Z*-double bond geometry. In biological systems, bilirubin shows potent antioxidant properties [85, 86] especially against peroxyl radicals [87].

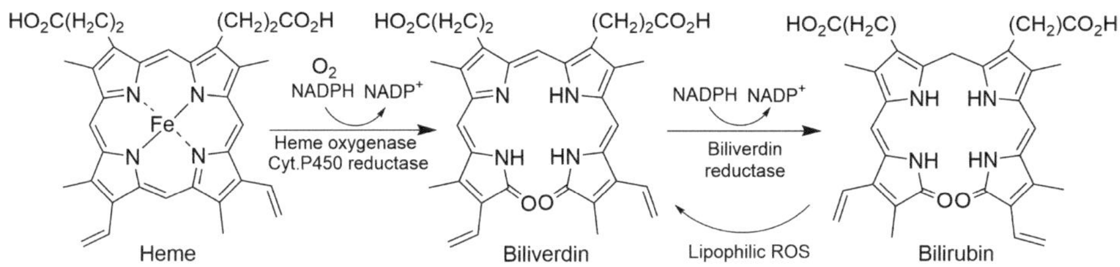

Figure 16.
Enzymatic degradation of hemoglobin heme to bilirubin.

6.5 Polyamines

Putrescine (H_2N-$(CH_2)_4$-NH_2), spermidine ([H_2N-$(CH_2)_3]_2$-NH), and spermine (H_2N-$(CH_2)_3$-NH-$(CH_2)_4$-NH-$(CH_2)_3$-NH_2) are biogenic unbranched polyamines (PAs) that exhibit antioxidant activities [88–90]. These amines are present in minute quantities in virtually all living species. While putrescine (1,4-diaminobutane) bears two primary amine groups at both terminal carbons, spermidine (triamine) and spermine (tetraamine) contain one and two additional secondary amine moieties, respectively. As antioxidants, PAs mediate protection of DNA against oxidative damage induced by hydrogen peroxide [90], scavenge free radicals [88], and reduce oxidative haemolysis of erythrocytes [90]. The amines also function as positive modulators of antioxidant genes under conditions of strong oxidative stress [88]. The protective effect of PAs is related to the stabilization of polyunsaturated phospholipids in cell membranes from peroxyl radicals, superoxides, and hydrogen peroxide [89]. In regard to their role in DNA protection against ROS, PAs are positively charged at physiological pH, enabling them to remain in proximity to negatively charged macromolecules, thus protecting them against oxidative damage [90]. Biosynthetically, the three polyamines are biosynthesized from L-ornithine, known to supply C_4N building block, and L-methionine [91]. In animals, L-ornithine undergoes a pyridoxal phosphate (PLP)-dependent decarboxylation to generate putrescine. Thereafter, aminopropylation of putrescine by the enzyme spermidine synthase and decarboxy-S-adenosyl methionine produces spermidine. Repetition of the same sequence of reactions in the presence of the enzyme spermine synthase generates spermine.

7. Conclusions

In addition to the oxidative damage that reactive oxygen and nitrogen species inflict on macromolecules, they also participate in damage caused by microbial infections, tumor progression, and neurodegenerative diseases. In response to such oxidative injuries, tissues protect themselves by expressing genes encoding antioxidant enzymes and endogenous antioxidants to maintain oxidants at harmless levels. Oxidants themselves mediate certain cellular functions and cannot be eliminated completely. This fact emphasizes the significance of the antioxidant defense system in maintaining homeostasis and normal physiological processes, and in combating diseases and promoting immunity. The regulation of gene expression by employing oxidants and antioxidants represents a novel approach with promising therapeutic implications. Exogenous antioxidants are also critical for maintaining healthy living and longevity and must be obtained through dietary means. However, excessive dietary supplementation may disrupt the activation of the endogenous antioxidant defense system. Consequently, further research is required to fully elucidate the importance of antioxidants in the therapy of several human disease states and promotion of health span.

Acknowledgements

Dr. Ziad Moussa is grateful to the United Arab Emirates University (UAEU) of Al-Ain and to the Research Office for supporting the research developed in his laboratory (Project 852).

Abbreviations

$AscH^-$	ascorbate
BIL	bilirubin
CAT	catalase
DHA	dehydroascorbate
DNA	deoxyribonucleic acid
DMAPP	dimethylallyl diphosphate
ESR	electron spin resonance
EnzSeSG	glutathione peroxidase selenenyl sulfide
EnzSeOH	glutathione peroxidase selenenic acid
EnzSeH	glutathione peroxidase selenol
GSH	glutathione
GSSG	glutathione disulfide
GPx or GSHPx	glutathione peroxidase
GR	glutathione reductase
GOx	glutathione oxidase
GSR	glutathione reductase
H_2O_2	hydrogen peroxide
HO·	hydroxyl radical
HClO	hypochlorous acid
IPP	isopentyl diphosphate
LDL	low-density lipoprotein
LOOH	lipid hydroperoxides
LOO·	lipid peroxyl radical
MPO	myeloperoxidase
NOS	nitric oxide synthase
NADPH	nicotinamide adenine dinucleotide phosphate
NO·	nitric oxide radical
ONOO−	peroxynitrite

$O_2.^-$	superoxide anion radical
PLP	pyridoxal phosphate
PAs	polyamines
PUFA	polyunsaturated fatty acids
PPP	phytyl pyrophosphate
ROS	reactive oxygen species
RNS	reactive nitrogen species
SOD	superoxide dismutase
SAM	S-adenosyl methionine
UA	uric acid
UDP	uridine diphosphate

Author details

Ziad Moussa[1*], Zaher M.A. Judeh[2] and Saleh A. Ahmed[3,4]

1 Department of Chemistry, College of Science, United Arab Emirates University, Al Ain, United Arab Emirates

2 School of Chemical and Biomedical Engineering, Nanyang Technological University, Singapore

3 Faculty of Applied Science, Department of Chemistry, Umm Al-Qura University, Makkah, Saudi Arabia

4 Faculty of Science, Department of Chemistry, Assiut University, Assiut, Egypt

*Address all correspondence to: zmoussa@uaeu.ac.ae

References

[1] Young IS, Woodside JV. Antioxidants in health and disease. Journal of Clinical Pathology. 2001;**54**:176-186

[2] We dinger A, Kosovo AV. Biological activities of reactive oxygen and nitrogen species: Oxidative stress versus signal transduction. Biomolecules. 2015;**5**:472-484

[3] Bagchi K, Puri S. Free radicals and antioxidants in health and disease. Eastern Mediterranean Health Journal. 1998;**4**:350-360

[4] Pisoschi AM, Pop A. The role of antioxidants in the chemistry of oxidative stress. European Journal of Medicinal Chemistry. 2015;**97**:55-74

[5] Łuszczewski A, Matyska-Piekarska E, Trefler J, Wawer I, Łacki J, Sliwinska-Stanczyk P. Reactive oxygen species-physiological and pathological function in the human body. Reumatologia. 2007;**45**:284-289

[6] Rosen H, Klebanoff SJ, Wang Y, Brot N, Heinecke JW, Fu X. Methionine oxidation contributes to bacterial killing by the myeloperoxidase system of neutrophils. Proceedings of the National Academy of Sciences. 2009;**106**:18686-18691

[7] Gęgotek A, Nikliński J, Žarković N, Žarković K, Waeg G, Łuczaj W, et al. Lipid mediators involved in the oxidative stress and antioxidant defence of human lung cancer cells. Redox Biology. 2016;**9**:210-219

[8] Csányi G, Miller FJ Jr. Oxidative stress in cardiovascular disease. International Journal of Molecular Sciences. 2014;**15**:6002-6008

[9] Morales-González JA, editor. Oxidative Stress and Chronic Degenerative Diseases—A Role for Antioxidants. Rijeka, Croatia: In-Tech; 2013

[10] McCord JM, Fridovich I. Superoxide dismutase: The first twenty years (1968-1988). Free Radical Biology and Medicine. 1998;**5**(5-6):363-369

[11] Epp O, Ladenstein R, Wendel A. The refined structure of the selenoenzyme glutathione peroxidase at 0.2-nm resolution. European Journal of Biochemistry/FEBS. 1983;**133**:51-69

[12] Deponte M. Glutathione catalysis and the reaction mechanisms of glutathione-dependent enzymes. Biochimica et Biophysica Acta. 2013;**1830**:3217-3266

[13] Chelikani P, Fita I, Loewen PC. Diversity of structures and properties among catalases. Cellular and Molecular Life Sciences. 2004;**61**:192-208

[14] Mirończuk-Chodakowska I, Witkowska AM, Zujko ME. Endogenous non-enzymatic antioxidants in the human body. Advances in Medical Sciences. 2018;**63**:68-78

[15] Nimse SB, Palb D. Free radicals, natural antioxidants, and their reaction mechanisms. RSC Advances. 2015;**5**:27986-28006

[16] Valko M, Rhodes CJ, Moncol J, Izakovic M, Mazur M. Free radicals, metals and antioxidants in oxidative stress-induced cancer. Chemico-Biological Interactions. 2006;**160**:1-40

[17] Matés JM. Toxicology, effects of antioxidant enzymes in the molecular control of reactive oxygen species toxicology. 2000;**153**:83-104

[18] Matés JM, Perez-Gomez C, De Castro IN. Antioxidant enzymes and human diseases. Clinical Biochemistry. 1999;**32**:595-603

[19] Knight JA. The aging process. In: Free Radicals, Antioxidants, Aging, and

Disease. Washington, DC: AACC Press; 1999. p. 64

[20] Rosen GM, Tsai P, Pou S. Mechanism of free-radical generation by nitric oxide synthase. Chemical Reviews. 2002;**102**(4):1191-1200

[21] Dizdaroglu M, Jaruga P. Mechanisms of free radical-induced damage to DNA. Free Radical Research. 2012;**46**:382-419

[22] Miaskiewicz K, Osman R. Theoretical study on the deoxyribose radicals formed by hydrogen abstraction. Journal of the American Chemical Society. 1994;**116**:232-238

[23] Pogozelski WK, Tullius TD. Oxidative strand scission of nucleic acids: Routes initiated by hydrogen abstraction from the sugar moiety. Chemical Reviews. 1998;**98**:1089-1108

[24] Steenken S, Telo JP, Novais HM, Candeias LP. One-electron reduction potentials of pyrimidine bases, nucleosides, and nucleotides in aqueous solution. Consequences for DNA redox chemistry. Journal of the American Chemical Society. 1992;**114**:4701-4709

[25] Candeias LP, Steenken S. Reaction of HO• with guanine derivatives in aqueous solution: Formation of two different redox-active OH-adduct radicals and their unimolecular transformation reactions. Properties of G(-H)•. Chemistry European Journal. 2000;**6**:475-484

[26] Dizdaroglu M. Oxidative damage to DNA in mammalian chromatin. Mutation Research. 1992;**275**:331-342

[27] Breen AP, Murphy JA. Reactions of oxyl radicals with DNA. Free Radical Biology and Medicine. 1995;**18**:1033-1077

[28] Esterbauer H, Schaur RJ, Zollner H. Chemistry and biochemistry of 4-hydroxynonenal, malonaldehyde and related aldehydes. Free Radical Biology and Medicine. 1991;**11**:81-121

[29] Bohm V, editor. Vitamin E. In: Antioxidants Special Issue. Basel, Switzerland: MDPI; 2017-2018. p. 1-148

[30] Hensley K, Benakass EJ, Bolli R. New perspectives on vitamin E: α-tocopherol and carboxyethylhydroxychroman metabolites in biology and medicine. Free Radical Biology and Medicine. 2004;**36**:1-15

[31] Hofius D, Sonnewald U. Vitamin E biosynthesis: Biochemistry meets cell biology. Trends in Plant Science. 2003;**8**:6-8

[32] Pryor WA. Vitamin E and heart disease: Basic science to clinical intervention trials. Free Radical Biology and Medicine. 2000;**28**:141-164

[33] Kojo S. Vitamin C: Basic metabolism and its function as an index of oxidative stress. Current Medicinal Chemistry. 2004;**11**:1041-1064

[34] Han R-M, Zhang JP, Skibsted LH. Reaction dynamics of flavonoids and carotenoids as antioxidants. Molecules. 2012;**17**:2140-2160

[35] Morales-Gonzalez JA, editor. Oxidative stress and chronic degenerative diseases: A role for antioxidants. In: The Exogenous Antioxidants. London, United Kingdom: IntechOpen; 2013. pp. 33-57

[36] Monaghan BR, Schmitt FO. The effects of carotene and of vitamin a on the oxidation of linoleic acid. The Journal of Biological Chemistry. 1932;**96**:387-395

[37] Parker RS. Absorption, metabolism, and transport of carotenoids. The FASEB Journal. 1996;**10**:542-551

[38] Vieira AV, Schneider WJ, Vieira PM. Retinoids: Transport, metabolism and mechanism of action. The Journal of Endocrinology. 1995;**146**:201-207

[39] Wolf G. The discovery of the visual function of vitamin A. The Journal of Nutrition. 2001;**131**:1647-1650

[40] Pehlivan FE. Vitamin C: An Antioxidant Agent. IntechOpen; 2017. DOI: 10.5772/intechopen.69660

[41] Mand J, Szarka A, Bánhegyi1 G. Vitamin C: Update on physiology and pharmacology. British Journal of Pharmacology. 2009;**157**:1097-1110

[42] Linster CL, Van Schaftingen E. Vitamin C. biosynthesis, recycling and degradation in mammals. The FEBS Journal. 2007;**274**:1-22

[43] Meister A. Glutathione-ascorbic acid antioxidant system in animals. The Journal of Biological Chemistry. 1994;**269**:9397-9400

[44] Noctor G, Foyer CH. Ascorbate and glutathione: Keeping active oxygen under control. Annual Review of Plant Physiology and Plant Molecular Biology. 1988;**49**:249-279

[45] Andersen OM, Markham KR, Flavonoids E. Chemistry, Biochemistry, and Applications. Boca Raton, FL, USA: CRC; 2006. pp. 1-1197

[46] Omar HS, El-Beshbishy HA, Moussa Z, Taha KF, Singab ANB. Antioxidant activity of *Artocarpus heterophyllus* Lam. (jack fruit) leaf extracts: Remarkable attenuations of hyperglycemia and hyperlipidemia in streptozotocin-diabetic rats. The Scientific World Journal. 2011;**11**:788-800

[47] Bravo L. Polyphenols: Chemistry, dietary sources, metabolism, and nutritional significance. Nutrition Reviews. 1998;**56**:317-333

[48] Morel I, Lescont G, Gogrel P, Sergent O, Pasdeloup N, Brissot P, et al. Antioxidant and iron chelating activities of the favonoids catechin, quercetin and diosmetin on iron-loading rat hepatocyte cultures. Biochemical Pharmacology. 1993;**45**:13-19

[49] Ratty AK, Das NP. Effects of favonoids on nonenzymatic lipid peroxidation: Structure activity relationship. Biochemical Medicine and Metabolic Biology. 1988;**39**:69-79

[50] Pekkarinen SS, Heinonen IM, Hopia AI. Flavonoids quercetin, myricetin, kaemferol and (+)-catechin as antioxidants in methyl linoleate. Journal of the Science of Food and Agriculture. 1999;**79**:499-506

[51] Stahl W, Sies H. Antioxidant activity of carotenoids. Molecular Aspects of Medicine. 2003;**24**:345-351

[52] Mueller L, Boehm V. Antioxidant activity of β-carotene compounds in different in vitro assays. Molecules. 2011;**16**:1055-1069

[53] Mortensen A, Skibsted LH, Truscott TG. The interaction of dietary carotenoids with radical species. Archives of Biochemistry and Biophysics. 2001;**385**:13-19

[54] Dewick PM. Medicinal Natural Products: A Biosynthetic Approach. 3rd ed. West Sussex, United Kingdom: John Wiley and Sons; 2009. pp. 298-300

[55] Andreasen MF, Landbo A, Christensen LP, Hansen A, Meyer AS. Antioxidant effects of phenolic rye (*Secale cereale* L.) extracts, monomeric hydroxycinnamates, and ferulic acid dehydrodimers on human low-density lipoproteins. Journal of Agricultural and Food Chemistry. 2001;**49**:4090-4096

[56] Teixeira J, Gaspar A, Garrido EM, Garrido J, Borges F. Hydroxycinnamic

acid antioxidants: An electrochemical overview. BioMed Research International. 2013:1-11

[57] Meyer AS, Donovan JL, Pearson DA, Waterhouse AL, Frankel EN. Fruit hydroxycinnamic acids inhibit human low-density lipoprotein oxidation in vitro. Journal of Agricultural and Food Chemistry. 1998;**46**:1783-1787

[58] Vaidya V, Ingold KU, Pratt DA. Garlic: Source of the ultimate antioxidants-sulfenic acids. Angewandte Chemie, International Edition. 2009;**48**:157-160

[59] Block E. The organosulfur chemistry of the genus allium—Implications for the organic chemistry of sulfur. Angewandte Chemie, International Edition. 1992;**31**:1135-1178

[60] Lynett PT, Butts K, Vaidya V, Garretta GE, Pratt DA. The mechanism of radical-trapping antioxidant activity of plant-derived thiosulfinates. Organic and Biomolecular Chemistry. 2011;**9**:3320-3330

[61] Amorati R, Lynett PT, Valgimigli L, Pratt DA. The reaction of sulfenic acids with peroxyl radicals: Insights into the radical-trapping antioxidant activity of plant-derived thiosulfinates. Chemistry–A European Journal. 2012;**18**:6370-6379

[62] Galano A, Francisco-Marquez M. Peroxyl-radical-scavenging activity of garlic: 2-propenesulfenic acid versus allicin. The Journal of Physical Chemistry. B. 2009;**113**:16077-16081

[63] Priyadarsini KI, Maity DK, Naik GH, Sudheer M, Unnikrishnan MK, Satav JG, et al. Role of phenolic O-H and methylene hydrogen on the free radical reactions and antioxidant activity of curcumin. Free Radical Biology and Medicine. 2003;**35**:475-484

[64] Salem M, Rohanib S, Gillies ER. Curcumin, a promising anti-cancer therapeutic: A review of its chemical properties, bioactivity and approaches to cancer cell delivery. RSC Advances. 2014;**4**:10815-10829

[65] Stinefelt B, Leonard SS, Blemings KP, Shi X, Klandorf H. Free radical scavenging, DNA protection, and inhibition of lipid peroxidation mediated by uric acid. Annals of Clinical and Laboratory Science. 2005;**35**:37-45

[66] Warning WS. Uric acid: An important antioxidant in acute Ischaemic stroke. QJM: An International Journal of Medicine. 2002;**95**:691-693

[67] Squadrito GL, Cueto R, Splenser AE, Valavanidis A, Zhang H, Uppu RM, et al. Reaction of uric acid with peroxynitrite and implications for the mechanism of neuroprotection by uric acid. Archives of Biochemistry and Biophysics. 2000;**376**:333-337

[68] Galano A, Alvarez-Idaboy JR. Glutathione: Mechanism and kinetics of its non-enzymatic defense action against free radicals. RSC Advances. 2011;**1**:1763-1771

[69] Settle T, Klandorf H. The role of uric acid as an antioxidant in neurodegenerative disease pathogenesis. Brain Disorders and Therapy. 2014;**3**:129-133

[70] Jiménez V, Alderete JB. Theoretical calculations on the tautomerism of uric acid in gas phase and aqueous solution. Journal of Molecular Structure (THEOCHEM). 2005;**755**(1-3):209-214

[71] Maples KR, Mason PR. Free radical metabolite of uric acid. The Journal of Biolocical Chemistry. 1988;**263**:1709-1712

[72] Lu SC. Glutathione synthesis. Biochimica et Biophysica Acta. 2013;**1830**:3143-3153

[73] Samuelsson M, Vainikka L, Öllinger K. Glutathione in the blood and cerebrospinal fluid: A study in healthy male volunteers. Neuropeptides. 2011;**45**:287-292

[74] Lushchak VI. Glutathione homeostasis and functions: Potential targets for medical interventions. Journal of Amino Acids. 2012;**2012**:1-26

[75] Rahman K. Studies on free radicals, antioxidants, and co-factors. Clinical Interventions in Aging. 2007;**2**:219-236

[76] Alli JA, Kehinde AO, Kosoko AM, Ademowo OG. Oxidative stress and reduced vitamins C and E levels are associated with multi-drug resistant tuberculosis. Journal of Tuberculosis Research. 2014;**2**:52-58

[77] Chatterjee A. Reduced glutathione: A radioprotector or a modulator of DNA-repair activity? Nutrients. 2013;**5**:525-542

[78] Flohé L. The glutathione peroxidase reaction: Molecular basis of the antioxidant function of selenium in mammals. Current Topics in Cellular Regulation. 1985;**27**:473-478

[79] Stadtman TC. Biosynthesis and function of selenocysteine-containing enzymes. The Journal of Biological Chemistry. 1991;**266**:16257-16260

[80] Dewick PM. Medicinal Natural Products: A Biosynthetic Approach. 3rd ed. London, United Kingdom: John Wiley and Sons; 2009. p. 366

[81] Laudon M, Frydman-Marom A. Therapeutic effects of melatonin receptor agonists on sleep and comorbid disorders. International Journal of Molecular Sciences. 2014;**15**:15924-15950

[82] Hardeland R. Antioxidative protection by melatonin: Multiplicity of mechanisms from radical detoxification to radical avoidance. Endocrine. 2005;**27**:119-130

[83] Reiter RJ, Acuña-Castroviejo D, Tan DX, Burkhardt S. Free radical-mediated molecular damage. Mechanisms for the protective actions of melatonin in the central nervous system. Annals of the New York Academy of Sciences. 2001;**939**:200-215

[84] Bonnefont-Rousselot D, Collin F. Melatonin: Action as antioxidant and potential applications in human disease and aging. Toxicology. 2010;**278**:55-67

[85] Zelenka J, Dvorák A, Alán L, Zadinová M, Haluzík M, Vítek L. Hyperbilirubinemia protects against aging-associated inflammation and metabolic deterioration. Oxidative Medicine and Cellular Longevity. 2016;**2016**:1-10

[86] Ziberna L, Martelanc M, Franko M, Passamonti S. Bilirubin is an endogenous antioxidant in human vascular endothelial cells. Scientific Reports. 2016;**6**:29240

[87] Hatfield GL, Barclay RC. Bilirubin as an antioxidant: Kinetic studies of the reaction of bilirubin with peroxyl radicals in solution, micelles, and lipid bilayers. Organic Letters. 2004;**6**:1539-1542

[88] Tkachenko AG, Akhova AV, Shumkov MS, Nesterova LY. Polyamines reduce oxidative stress in *Escherichia coli* cells exposed to bactericidal antibiotics. Research in Microbiology. 2012;**163**:83-91

[89] Fujisawa S, Kadoma Y. Kinetic evaluation of polyamines as radical scavengers. Anticancer Research. 2005;**25**:965-969

[90] Rider JE, Hacker A, Mackintosh CA, Pegg AE, Woster PM, Casero RA Jr. Spermine and spermidine mediate protection against oxidative damage caused by hydrogen peroxide. Amino Acids. 2007;**33**:231-240

[91] Dewick PM. Medicinal Natural Products: A Biosynthetic Approach. 3rd ed. London, United Kingdom: John Wiley and Sons; 2009. pp. 311-313